Taking the Back off the Watch

For further volumes:
http://www.springer.com/series/5664

© Carvel Gold

Thomas Gold

Edited by
Simon Mitton

Taking the Back off the Watch

A Personal Memoir

Advancing Astronomy and Geophysics

Springer

Author
Thomas Gold
Cornell University
Ithaca
New York
United States

Editor
Simon Mitton
St. Edmund's College
Cambridge
United Kingdom

ISBN 978-3-642-27587-6 ISBN 978-3-642-27588-3 (eBook)
DOI 10.1007/978-3-642-27588-3
Springer Heidelberg New York Dordrecht London

Library of Congress Control Number: 2012938100

Printed on acid-free paper

Springer is part of Springer Science+Business Media (www.springer.com)

Foreword

In the year 1968, astronomers began to change the way they think about the universe. The change was deep and fundamental. Before 1968, the universe was seen as a peaceful assemblage of slowly evolving objects, and the job of an astronomer was to construct an accurate picture of an almost unchanging landscape. After 1968, the universe was seen as a drama full of violent events and cataclysms, and the job of an astronomer was to observe and understand the processes of change. The shift in the culture of astronomy took many years to complete, but the year 1968 was a crucial turning point. Two people were responsible for the sudden shift, Jocelyn Bell and Tommy Gold. Jocelyn Bell, a young radio astronomer in England, discovered the rapidly pulsating radio sources that later became known as pulsars. Tommy understood what the pulsars were and why they were important.

I remember vividly the events of 1968. In February, we heard the news of Jocelyn Bell's discovery. Everyone who was interested in astronomy was trying to understand what these mysterious radio pulses could mean. At that time, we knew that there were two kinds of stars. There were normal stars, made of hot gas at normal density, like our sun and all the other stars that we can see with our naked eyes. There were white-dwarf stars, made of gravitationally collapsed matter, with densities about a million times greater. In addition to these two well-known kinds of stars, there was some talk of a possible third kind called neutron stars. Nobody had ever seen a neutron star. Neutron stars were a theoretical idea originally invented by Fritz Zwicky in 1931. Zwicky was a physicist and not an astronomer. For that reason, few astronomers took his idea seriously. Zwicky's idea was that neutron stars should have densities about a million times greater than white dwarfs. A neutron star with the mass of our sun would be only about ten kilometers in size. Even if neutron stars existed, they would be so small that they could never be observed, and they were therefore of no interest to traditional astronomers.

Like everyone else in 1968, I knew that normal stars sometimes pulsate, becoming brighter and fainter in a regular cycle with periods of a few hours or days. These pulsating normal stars are called Cepheid variables and are well understood. The period of a pulsating star is roughly the time it takes for a sound wave to travel across the star and back again. The pulsation is an unstable sound wave that resonates with high amplitude, like the air in a loudly played trombone. It was then easy to calculate that a typical white dwarf, with a million times greater density and a greater sound velocity, would have a pulsation period of the order of a second. Jocelyn Bell's first pulsar had a period of 1.337 s. Obviously, it was a pulsating white dwarf. I believed that. Everyone else believed it too, except for Tommy. The problem remained, how to understand why a pulsating white dwarf should emit enormously powerful pulses of radio waves for our entertainment. In May 1968, the first International Conference on Pulsars was held in New York. I was there and heard the experts trying to explain the radio pulses. They all assumed, like me, that the source was a pulsating white dwarf. Tommy was there too. He was not invited to speak, since the organizers of the meeting considered his ideas to be obviously crazy. But he spoke anyway and told us the true explanation of pulsars.

Tommy had been thinking about neutron stars long before pulsars were discovered. He knew that if neutron stars were born as Zwicky had suggested, when normal stars collapse in supernova explosions, the neutron stars are likely to be born with extremely fast rotation and extremely strong magnetic fields. With fast rotation and strong fields, the neutron star could emit radio signals like a lighthouse with a rotating beam. The rotating radio beam would look like a pulsar. As soon as Tommy heard of Jocelyn Bell's discovery, he knew that this was the evidence he had been waiting for, proving that neutron stars really exist. He told us at New York that neutron stars should often be born spinning even more rapidly than the one that Jocelyn Bell had found, with periods much shorter than a second. If his explanation was right, pulsars should exist with periods measured in milliseconds. If pulsars were white dwarfs, then periods much shorter than a second would be impossible. In November of the same year, as a result of Tommy's suggestion, a pulsar with period 33 ms was discovered at the exact place in the sky where a supernova explosion had been observed by Chinese and Korean astronomers in the year 1054. This millisecond pulsar immediately demolished the white-dwarf theories and proved that Tommy's theory of pulsars was right. Neutron stars were real and abundant in the universe. They would not only be producing radio pulses but would have many other observable effects. They would be spewing out vast quantities of magnetized plasma and high-energy particles into the galaxy. In one year, our vision of the universe was transformed.

Besides explaining pulsars and proving that neutron stars exist, Tommy left his imprint on the world of astronomy in other ways. In 1960, he moved from Harvard to Cornell University, where he became chairman of the astronomy department with authority to expand the department and to create a brand-new "Center for Radiophysics and Space Research." Under his management, the Center for Radiophysics became a world-class institution for people doing nontraditional kinds of astronomy. One of the luminaries whom Tommy attracted to the center was Carl Sagan. Sagan became firmly attached to Cornell and stayed there for the rest of his life. Sagan was like Tommy, interested in everything and always ready to plunge into new ventures. The two of them were a good team, Sagan as the spellbinding television personality who made things in the sky exciting to the public, Tommy as the organizer who made things happen on the ground.

One of the things that Tommy made happen was the conversion of the Arecibo Radio Telescope into a superb instrument for radio astronomy. The Arecibo telescope had been built by the US Department of Defense as an instrument for studying the ionosphere. The ionosphere is important for military communications and is subject to rapid fluctuations. The telescope was designed as a powerful radar giving prompt information about the current state of the ionosphere. By good luck, Cornell University was responsible for running the telescope. When the telescope was built in 1963, Tommy was in charge of its nonmilitary operations. He quickly found out that the telescope worked very well for the ionospheric observations for which it was designed but worked very poorly for radio astronomical observations which require high sensitivity. Without high sensitivity, the telescope could not take advantage of its huge size to see faint radio sources at the edge of the universe. It turned out that the correction of this design flaw required a complete replacement of large parts of the instrument. The replacement was costly in time and money. Tommy had to fight hard to get it done. It took nine years before it was finished. Without his stubborn determination, it would probably not have happened. In the end, after the nine years were over, the Arecibo telescope became what Tommy had intended it to be, the finest instrument in the world for studying faint point sources such as pulsars.

Tommy loved Arecibo and was intensely proud of what he had done there. On one stormy day in Puerto Rico, he was acting as tour guide to a visiting committee of scientists who came to Arecibo to see the telescope in operation. He led us along a high catwalk, starting on one of the mountains from which the antenna is suspended and ending at the platform of the antenna. We emerged on the platform, a huge steel structure, suspended 500 ft above the reflecting dish on the ground below us. The platform was steady as a rock, with no perceptible motion, although the wind was whistling in our ears and lightning flashes were visible from an approaching storm. Tommy walked ahead of us to the edge of the platform, which was unprotected by any kind of railing. He chose this moment to give us a lecture, describing the history of the telescope and the plans for its future. He talked with great animation for 20 min, facing us with his heels on the edge of the platform, with nothing behind him but a 500-foot drop into thin air. We were huddled together in front of him, nervously waiting for the gust of wind that would blow him into oblivion, trying to listen to what he was saying. At the end of 20 min, he calmly asked for questions and was in no hurry to move away from the edge. That was Tommy, a great circus performer as well as a great scientist, proud of his physical mastery as well as of his intellectual achievements. The Arecibo telescope remained at the forefront of radio astronomy for 40 years and is only now beginning to become obsolete. It is not only a wonderful scientific tool but also a wonderful piece of architecture, perhaps the most beautiful man-made object of the modern era. It is a fitting monument to the life and work of Tommy Gold.

In the last years of his life, Tommy's main interest was the origin of hydrocarbons on the Earth. The hydrocarbons are natural gas and oil, which also happen to be of immense importance to the economic activities of humans and to the ecology of the planet. The prevailing view of experts in Western countries is that hydrocarbons are by-products of biology, resulting from the decomposition of bodies of ancient plants and microbes that died and were buried for millions of years. Tommy, together with many of the petroleum experts in Russia, took the opposite view, believing that the hydrocarbons came up from nonbiological sources deep in the Earth and only became contaminated with biological molecules because microbes invaded and fed on them on their way up. It was natural for an astronomer to prefer a nonbiological origin of hydrocarbons, since hydrocarbons are abundant in the outer planets of the solar system and were probably abundant in the material out of which the Earth condensed. Tommy's views are still fiercely opposed by a majority of experts. The debate rages on. I have to confess that I am like Tommy, more familiar with astronomy than with petroleum chemistry, and I believe that the experts are wrong.

Tommy was tough and never gave up a fight. He liked to tell us how he learned to be tough, as a Jewish kid fighting gangs of Nazi hoodlums in the streets of Berlin in the years 1930–1933 when Hitler was rising to power. He was a street kid from age 10 to age 13. The hoodlums knew that he was Jewish. They were mostly much bigger and older, but he was quicker and smarter. He learned to hit them hard and never show fear. Those years in Berlin shaped his character for the rest of his life.

As a final glimpse of Tommy, I like to recall a morning in the Swiss ski resort of Arosa in the spring of 1947. There is an almost vertical mountainside overlooking Arosa. It was then deeply covered with fresh snow. It had one or two steeply winding trails, for expert skiers only. That morning, the whole town was astonished to see four parallel pairs of ski tracks coming straight down the mountain from top to bottom, avoiding the trails and jumping over minor obstacles such as rocks and cliffs. Everyone was wondering who the four mysterious strangers could be. The four strangers had evidently come down the mountain together at high speed like dive bombers. Only a few of us who knew Tommy could guess what had happened. Tommy had gone out by himself before breakfast and skied down the mountain alone four times, each time keeping at a fixed distance from his previous track. That was his way of showing the world what he could do.

NJ, USA Freeman Dyson

Introduction

Thomas Gold (1920–2004) was one of the most remarkable astronomers in the second half of the twentieth century. In a career that spanned half a century, he worked in astrophysics, cosmology, physiology, radio astronomy, geophysics, and lunar science, where he was by turns an innovative practitioner and a daring theorist. After completing his war service in naval intelligence, he held positions at the University of Cambridge, the Royal Greenwich Observatory, Harvard University, and finally Cornell University in 1959.

Gold was born in Vienna on May 22, 1920, where his father was a wealthy industrialist with the means to provide a comfortable and privileged life for his son and heir. However, the economic crisis of the late 1920s meant the family moved to Berlin where his father became a metals trader. Gold's father, Max, was of Jewish origin from the Polish Ukraine and his mother was of German Catholic origin. Neither parent had the slightest interest in religion. They fled Germany in 1933 and settled in England in 1937. His father's gift of a watch, which Gold took apart and reassembled, led to his interest in technology, and Thomas entered Trinity College, Cambridge, in October 1939 to read engineering.

In May 1940, the British Government introduced the internment of all men of German or Austrian descent who were resident in eastern England. Behind the barbed wire, Tommy encountered Hermann Bondi, who had fled from Vienna, and this was the start of a lifelong friendship. The internees were transferred to Canada, and then brought back to England in 1941. Gold completed his engineering degree, getting a miserable result, an ordinary degree, in 1942. Given the scale of what he later achieved, the poor result can be put down to the fact that almost all of the teaching faculty at Cambridge were engaged elsewhere on the war effort, with instruction left in the hands of elderly dons.

Those who were away on war work included the Cambridge theoretical physicist Fred Hoyle who was at a secret experimental naval radar station from May 1940. By 1942, Hoyle was in charge of the theory group at the Admiralty Signals Establishment, where Bondi and Gold joined him. Thus was the stage set for a dazzling collaboration of cosmologists. The trio had many discussions about an unsolved problem in cosmology that had surfaced in the 1930s: the age of the Earth appeared to be twice the age of the universe. Gold was the first to propose, in late 1947, what came to be known as the steady-state theory of cosmology. Hoyle and Bondi, who were more accomplished applied mathematicians than Gold, worked through the implications of Gold's natural philosophy that the universe preserves its appearances throughout time, existing in a stable state of expansion with continuous creation of new matter. In 1948, Gold and Bondi published a paper on the philosophical aspects of the new theory for the origin of the universe, while Hoyle published his theory of a creation field to account for the new matter compensated for the effects of expansion. In the eyes of the general public, the theory received a massive boost as a result of Hoyle's BBC radio broadcasts in 1949 and 1950, in the first of which he proposed the name "big bang" to describe the rival theory. The discovery of the cosmic microwave background in 1963 led to general acceptance of the big bang theory, although Gold never lost faith in steady-state cosmology.

In 1949, Gold was appointed to a junior faculty position in the Cavendish Laboratory (Department of Physics) at Cambridge. This brought him into daily contact with Martin Ryle in the radio group at the laboratory. In those days, Ryle's group were doing fundamental development work on instrumentation for investigating cosmic radio sources, of which some 50 were known. Gold was sucked into furious arguments between Ryle and Hoyle, who were both slightly older than him. Ryle felt that the cosmic sources were stars in our galaxy, whereas Hoyle and Gold believed them to be extragalactic. Gold unwisely engaged in loud criticism of Ryle and that led to his modest appointment not being renewed.

His luck changed remarkably when the astronomer royal offered him the post of chief assistant (a senior position despite its title) at the Royal Greenwich Observatory. Gold was given a free hand, and he decided his future was in space research. He took the first steps in areas that would become mainstream in his professional career: the nature of the lunar surface and the role of magnetic fields in space. The Oxford English Dictionary gives him credit for coining the noun magnetosphere:

> *1959 T. Gold in Jrnl. Geophysical Res. 64 1219/1 The region above the ionosphere in which the magnetic field of the earth has a dominant control over the motions of gas and fast charged particles ... is known to extend out to a distance of the order of 10 earth radii; it may appropriately be called the magnetosphere.*

He had to resign abruptly from the Royal Observatory in 1956 because he was unable to work with the newly appointed astronomer royal, Richard Woolley. That is because Woolley was implacably opposed to space research and radio astronomy, the very fields in which Gold would excel. He and his family left England for good, accepting a chair in radio astronomy at Harvard in 1957. Two years later, Cornell University offered him the headship of its department of astronomy, and Gold made his final move.

At Cornell, he founded and directed the Center for Radio Physics and Space Research. Then he persuaded the US Defense Department to fund the construction of the giant radio telescope at Arecibo in Puerto Rico. Commissioning the telescope proved to be immensely challenging, but Gold overcame shortcomings of the design and went on to make great discoveries. This period of Gold's career is recounted in Chap. 7.

During the preparations for the lunar landings, Gold courted controversy with NASA over his assertion that the lunar dust could be several meters thick. As Gold recalls in Chap. 9, eventually, relations improved and he served with distinction on NASA's science advisory committee. He was opposed to the manned space flight programs and the development of the shuttle on the grounds that robotic instruments could deliver the science at much lower cost. In this, he turned out to be completely correct.

In 1968, the discovery of pulsars by the Cambridge radio astronomy group excited worldwide interest. I was a graduate student in this group, and I remember everyone's astonishment when Gold suggested that pulsars were rotating neutron stars. He concluded that radio astronomers should be able to detect shorter duration pulses. The Arecibo telescope did indeed find a pulsar with a period of 0.033 s in the Crab nebula. Chapter 8 is his exciting account of his role in the pulsar era.

From 1980, he became fascinated by the origin of petroleum. He believed that methane was primordial rather than biogenic. Hydrocarbons are common in the gas clouds that are star formation regions. Gold believed that what came to be known as fossil fuels were in fact the end result of material that was present inside the Earth from the formation of our planet in the solar nebula.

His honors included being made a Fellow of the Royal Society (1964), Fellow of the American Academy of Arts and Sciences (1974), election to the National Academy of Sciences (1974), and the award of the gold medal of the Royal Astronomical Society. He died on June 22, 2004, at Cayuga Medical Center, Ithaca, New York.

Gold had almost finished the preparation of these memoirs when he passed away. I learned of the existence of this memoir from Hermann Bondi shortly before he died on September 10, 2005. I had interviewed Hermann a few times in order to get material for my biography: *Fred Hoyle, a life in science* (Cambridge 2011). Bondi told me that Carvel Gold, Tommy's second wife, was having difficulty in finding a publisher. In October 2008, I was at Cornell University for a planetary science meeting. This gave me the opportunity to view the typescript. As an academic historian of science, I was personally motivated to edit the autobiography and find a publisher for it. For some time, it seemed that publishers were taking the view that the moment had passed for publication because it would be difficult to market an autobiography of a deceased scientist. Fortunately, I was able to secure publication within Springer's niche of history of astronomy books published in the series "Astrophysics and Space Science Library," resulting in the inclusion of this book in the joint RAS-Springer book series.

Perhaps I can be allowed to say a few words on the editorial process since this may be relevant to scholars and historians of science who wish to use the memoir as a source book, as I hope many researchers will. Gold's typescript was sent to three eminent research scientists for peer review. The reviewers unanimously supported publication and made valuable suggestions about how to improve the narrative. The reviews were the key to getting the support of Springer. When I commenced my work on editing and developing the typescript, I set myself a number of rules. I have added nothing to the material that Tommy left. On the advice of reviewers, I made cuts of about 15% in order to tighten the narrative flow. Generally, I deleted material that was of marginal interest. I have made a significant number of minor corrections to factual and historical material. Also, I have made grammatical and stylistic changes in order to improve the prose. These are changes that a competent copyeditor would have required. In the later chapters, I occasionally toned down some of the critical comments because I felt they had potential to cause offense. Again, I think any publisher averse to the risk of legal action would have required these alterations.

Finally, I wish to express my profound thanks to Carvel Gold for allowing me to edit this account for publication. Carrie's enthusiasm for this project got me through some difficult moments, and she always responded immediately to my many requests for advice.

Cambridge, UK Simon Mitton

Prologue

When I was 11 years old, my father gave me my first watch. It was of course the ticktock variety of a wristwatch, quite expensive in those days. The next day, he asked me why I did not have it on. I made some evasive reply, but the question was repeated the next day and the day after. Finally, I had to come clean and admit that I was putting it together. "How had it come apart?" "I had to see what was inside."

Its minutest constituents were in fact all lying on my desk. I had bought a watchmaker's screwdriver and tweezers and had taken the watch apart very carefully, placing all the little wheels on my table exactly in the pattern in which they had been in the watch. But when I uncaged the mainspring, it exploded, and all components were scattered around the room. Now, it was going to be much harder to assemble it again if I had to see where each gearwheel would fit into the housing. When my father saw what I was doing, he told me to use my allowance and take all the pieces to a watchmaker who might be able to assemble it. But I took this as a challenge and worked even harder at it. I do not know how many times I took the internal housing apart again to test the position of each wheel, but finally, I had it together—and it worked! With great pride I showed it to my father, but he was skeptical, and for good reason: lying on my desk was another little gearwheel that I had not fitted in. All I could say was that evidently this gearwheel was superfluous, since the watch worked fine without it. It did, and I had it for many years, but never solved the mystery of the extra gearwheel.

"Getting the back off the watch to see what is inside" was pretty much my philosophy throughout my scientific life. I tried to peer into the structure and behavior of the universe, the function of human sense organs, the properties of sun and moon and the solar system and our own little planet, processes that go on in stars of unimaginably high density, and numerous other things that puzzled me.

I believe that a scientist is a person who has never settled down and accepted the world as it is, but one who keeps marveling with childish curiosity at anything he cannot understand. No other motivation has produced such good science as this: not direction by government or funding agencies, nor by any other superior authority, nor promise of personal gain or acclaim.

Contents

Chapter 1
School Years in Troubled Times

Family Background

I was born in Vienna in 1920, shortly after the end of World War I, and I lived the first 10 years of my life there. Needless to say, I understood very little of what was going on in the world, but perhaps the errors in the interpretations that I made of the world around me in those years are more important to record than what I now understand. I lived with my mother, father, and 5-year-older sister in a large house on the outskirts of Vienna, and whatever went on there was regarded by me in my young years as the normal goings on in a family. My father was picked up punctually every morning by our chauffeur and driven to his office, where he apparently directed a very large organization. I did not know too much of what this meant. I would usually have breakfast with him, but after breakfast, I would go upstairs to the kitchen where the nice cook, a lady from Bohemia (eine böhmische Köchin), supplied me with whatever goodies she could either find in the larder or make up for me. She was a marvelous cook, and I still remember many of her dishes. Meanwhile, my mother, who would usually still be in bed, would hand out to the maid a list of groceries to be obtained on her rounds to the butcher, and the baker, and the greengrocer, and whatever other stores had to be visited to complete the list. (Supermarkets were not introduced in Europe until much later.) Then I would probably go up to my nanny, much of the time a nanny from France, engaged to teach me French. Although my French vocabulary remained very limited, the French pronunciation that I learned has largely stayed with me and helped me in later years. Then there were the piano lessons and the never-ending études that I had to practice for the next day. Or I would go and play around in our spacious garden or help the gardener with some of his jobs.

To me, this was the normal life of a family: a chauffeur, a maid, a cook, a footman, and a nanny. My view of this was not changed much by all the other families that I came to know from the frequent dinner parties, sometimes of as many as 12 persons. The stories from their households that I heard were quite similar to ours.

Thomas Gold *edited by* Simon Mitton, *Taking the Back off the Watch*,
Astrophysics and Space Science Library 381, DOI 10.1007/978-3-642-27588-3_1,

I remember once being taken to visit my father in his office in downtown Vienna overlooking one of the well-known Viennese squares; he was sitting behind a very large desk in a rather imposing room. Many other persons seemed to be engaged in working in other rooms that seemed to belong to the same organization. So that is what fathers do when they are directors of large organizations, was my only thought at this stage. When I grew up I would one day be the director of a large organization, and that seemed inevitable.

What I discovered much later was that not all families had servants or a father in the position of head of a large company. It only dawned on me then how really extraordinary my position had been. My father had been one of five brothers: in order of ages they were Alfred, Oscar, Emil, Leo, Max, with my father, Max, the youngest. They were of Jewish origin having immigrated to Vienna from the Polish Ukraine. My mother of German Catholic origin had been a child star of considerable fame on the Vienna theater stage, as I could see from the volumes of newspaper articles she had saved describing her performances. My father had been a lieutenant in the Austrian Army in World War I, that great war, forever remembered by the enormous casualties suffered on both sides but not remembered for any purpose other than possibly to redefine the boundary lines between different countries (Fig. 1.1).

Of my father's brothers, Emil had died very young, and we knew little about him. But of the others, all seemed to have had very successful careers in post-World War I Vienna. Alfred had become an art expert, a consultant concerning valuable pictures, and a dealer in such goods. His family also had a special interest for me, namely that Alfred's daughter, Marianne, or called Mecki in the family, was of all

Fig. 1.1 In this studio photograph taken during World War I, Gold's father Max is front left in his uniform as a lieutenant in the Austrian Army

the many cousins I had, by far my favorite one. She had not only charm and beauty but was also a person of wit and humor, sparkling eyes, and infectious laughter. She spoke several languages fluently; she had obtained a degree in chemistry and in sculpture from the University of Paris. From my earliest years, I was very keen to see and hear her. She was about 15 years older than me, and in the years in which I often saw her, she was already an accomplished sculptor. Also, it was a pleasure to watch her, pencil and paper in hand, to sketch whatever she happened to be looking at. Mecki was married to Mr. Littman, and her name therefore, on much of her work, is Marianne Littman. She had later a very successful career, mainly on the West Coast of the United States in Portland, Oregon, where she taught sculpture at the University and also contributed many sculptures to public and other major buildings. Later, she was divorced from Mr. Littman in circumstances we do not know, but eventually, when she found out that her ex-husband was seriously ill, she joined him again just to help looking after him.

The next of my father's brothers, Oscar, became a leading attorney in Vienna.

Leo, perhaps the most remarkable personality in the whole group, became a leading contender in world class chess. On the side, he had accepted the position of head of the Austrian branch of one of the global mining corporations. In that position, he had become wealthy very quickly and could retire early to devote himself entirely to chess. He moved his residence from Vienna to Zurich, and since he had married a person whose cooking or home keeping he did not appreciate, he decided to settle in an apartment serviced fully by the Dolder Hotel in Zurich, at that time known as one of the best European hotels.

My father, the youngest, had studied law at the University of Vienna, and from the lowly position of a lieutenant in the Austrian army, he had quickly moved up to becoming the director of Austria's largest metal refining and mining corporation.

These were the years in which Vienna was at a high point in intellectual activities. It was the time when Vienna was leading the world in medicine, in music, including the performances at the Vienna opera, in literature, in design, and in architecture. I eventually learned that my father had not just been the hardworking businessman that I knew him to be, but he was also a member of discussion groups that included many persons whose names we now know from this period.

My father had very strict views on how an immigrant should behave in the country to which he had emigrated. "Never try and stand out in any way either by attire, appearance, or behavior. Get used to the way people behave in the country in which you have chosen to live." Religion was no problem for him since he had in any case no interest in this subject at all. The same rules were shared by his other brothers who had all immigrated with their parents to Austria. They all spoke High German with precision and not the common Austrian dialect which of course was spoken by our employees as well as by the taxi drivers and by most of the people with whom one came into contact in the streets. Unlike Switzerland, where the local dialect became the choice of people who wanted to emphasize their patriotism, in Austria, High German was the language of choice, and the Austrian dialect was only spoken by those who were brought up entirely with that language and had probably never learned to speak High German. Dialects divide the German-speaking countries very sharply so that a person from North Germany such as Berlin or Hamburg could really not understand a Viennese taxi driver. In my four school years in Vienna, I brought home many words of the Austrian dialect and I was sharply pulled up by my father whenever I used them. In fact, I did not know anything about the background of my father or his family since he never brought it up and his parents both had died before I was born.

Schooling in Berlin

In 1930, when I was 10, the comfortable life in Vienna suddenly came to an end. I assume that the decision of my father to accept the directorship of a company based in Berlin and resign from the Vienna company was connected with the great worldwide depression. But as a child, I understood nothing about that. At any rate, we sold our lovely house in Vienna and bought another, somewhat more modest one in Berlin. I had to accommodate to that change and also to see more of the real world.

I remember little about my 4 years in the Vienna school, but it was certainly not unpleasant. The first part of my schooling that I can remember clearly was the new school in Berlin that I attended for 3 years from 1930 to 1933. These were of course the dreadful years when the German Republic was coming apart, street fighting raged between the Nazi storm troopers of Hitler and the opposing socialists and communists, and when anti-Semitism was becoming rampant. We were living in a quiet little street away from all the turbulence, but the way to and from school was not always so quiet. I disliked the school intensely; it was a very large school with classes of 50 or more, with many ruffians and young thugs who presumably had taken their ways from the rapidly decaying morality of the adults. Fights between the boys, some quite serious, were a common occurrence. My schoolwork was poor, and in the German system, it was always a problem to reach the level at which one would be transferred into the next class at the end of the school year. There were some in my class of 50 boys who failed to get through and had to repeat the year. I regularly had the lowest grades of those allowed to move on. As I heard much later, the teachers contacted my parents and suggested to them that I was possibly mentally retarded.

In gymnastics, which was treated as a serious subject, I was, however, all the time one of the best. I was small for my age, but as far as I can remember, in fights, I could defend myself very well against some of the biggest boys in the class. Street fights were a common, almost daily, occurrence.

Some of the Nazi thugs in the class had identified my name as a Jewish one, and the Jews were commonly thought to be physically weaker than the "Herrenfolk," the German race. I think the fact that I was a tough street fighter irked them particularly. There would often be three or four of the thugs waiting around some street corner for me and with shouts of "Judenbube" (Jewboy) they would pounce upon me and try and beat me up. I could not imagine any better training for street fighting or for educating me to despise prejudice. We will come to an episode later, where it seems likely that this training saved my life.

During classes, it was my habit to sit quietly in the back row so that the teacher's voice would not disturb me too much, as I was reading cheap detective stories on my lap, concealed under the desk. Sometimes I was so engrossed in the detective story that I did not even know what subject was being taught at the time. I suppose this helped me to become a good and rapid reader, but other than that, I learned practically nothing.

In my last year in the Berlin school, when I was 13, one of the subjects taught was Euclid's geometry. There it seems I suddenly began to listen and occasionally abandon my book for 5 or 10 min to hear about this or that proof of Euclid: triangles, quadrilaterals, inscribed and circumscribed circles, and all the proofs that were fundamental to these matters, many of which seemed quite obvious to me. Then, on one occasion which I remember very clearly, I was suddenly called to the board to present a proof of some theorem, one of many we were supposed to have studied for ourselves in the book. Of course I had not done any studying, and I am sure the teacher expected to send me back and ask one of the better students to take over. However, as I looked at the problem, I saw immediately a way of proving the theorem and proceeded to do so. "Where does that proof come from?" the teacher asked with surprise, for evidently, it was not the proof given in the schoolbook. "I don't know," I said, "I just made it up." "Remarkable," said the teacher, "Yes, your proof is perfectly correct, but we had better have someone explain the proof that is in the book." As I learned later from my parents, the teacher got in touch with them and said that they had evidently been quite wrong in regarding my poor performance as due to a mental handicap; instead, it now seemed to them that I was really quite bright but extremely lazy. They would therefore have to be much tougher in dealing with me.

Escape from Berlin

Luckily for me, I was not subject to this toughness for long, because as Hitler had come to power, my father decided that it would not be safe for us to stay in Berlin, and he prepared to send me to a boarding school in Switzerland. We were Austrian citizens, and at that time, this still shielded us from confrontation by the Nazis. However, when one day the government proclaimed that it had directed the police and the storm troopers to "deal with the Jews," of course all hell broke loose and my father decided that the family should leave Germany immediately. The drive to the railway station gave a clear warning of what was to come. As we drove through some of the major streets, we saw many shop windows smashed. People were being dragged out into the street and beaten by police, by Nazi storm troopers noted for their brutality and violence, and also by many members of the public. The Nazi mentality had really asserted itself.

We got to the railway station where we met up with a good friend of my father's, a wealthy German-Jewish businessman. He had the strongest reason to leave in a hurry and, of course, had no intention of ever returning to Germany. As I learned later from my father, this businessman had spent the first part of the day converting as much money as possible into very compact items of value such as diamonds, expensive jewelry, and high-denomination banknotes. He told my father, once we were under way, that he had hidden a little bag with all these precious objects and that he had found such a subtle hiding place on the train that he thought that even the most thorough search, such as might be undertaken at the border, would be unlikely to reveal it. He did not tell my father where he had hidden it, saying that he knew that he was subjecting himself to a very serious risk, and he wanted to make sure that there was no way in which my father might get involved. The train was taking us to Vienna by the normal route, which goes through Czechoslovakia. When the train stopped at the German–Czech border, a large number of uniformed men came on board, looked everyone over, scrutinized passports and identity papers, and asked questions. Then, quite arbitrarily it seemed, they took some people off the train for further questioning, and our friend was among them. On the way from the German border to Prague, my father tried desperately, of course, to retrieve that little bag, and he searched all the places he could think of. He did not find it. In Prague, we had to change trains, and the last thing we saw of the train with the goods on board was that it was shunted onto some sidetrack. We went on to Vienna where we heard from our friend who had in fact been released and had taken the next train to Prague. There he spent several days going through the railway yards trying to find out where the particular carriage had gone. He could not find out, and he arrived in Vienna with the realization that he was now no longer wealthy but in fact extremely poor. His bank accounts and safe deposit boxes in Berlin had all been taken over by the Nazis, and he certainly did not dare to return to try and recover any of this. I think my father supported him for a while and so did other friends he had outside Germany. I suppose that a cleaning lady or a train mechanic in Czechoslovakia had suddenly become very wealthy. This story had a sad ending. Our friend moved to the Netherlands, and when the Nazis overran that country in 1940, he was captured and died in a concentration camp.

School in Switzerland

The change from Berlin to the boarding school in Switzerland was unbelievable. I suppose I had no conception at the time that the world contained such contrasts. The school was situated in Zuoz, a small village in beautiful surroundings, in a high mountain valley, the Engadine. Everyone was friendly to me from the start. What a difference from the nastiness and the dirty squalor of Berlin! My father took me there. He worried that I might feel lonely and abandoned, the first time away from home at the age of 13. There was no such problem. I loved the school, and I was very happy to be there. The headmaster went out of his way to make new boys feel at home, and he saw to it that I was guided by boys who knew the ways of the school and who he judged to be appropriate companions. I soon had many friends (Figs. 1.2–1.4).

The center of the village of Zuoz was an open square, surrounded by magnificent old buildings, and including a few shops and restaurants. The main point of attraction for us was the "Konditorei," or "Konda," as we called it, where superb bakery products were always available in the best tradition of Swiss baking. We would enjoy such delicacies as Marzipankartoffeln (marzipan "potatoes"),

Fig. 1.2 The village of Zuoz, Switzerland, in February 1951. The Lyceum Alpinum Zuoz is an international boarding school founded in 1904

Fig. 1.3 Young Tommy aged about 13 skiing on a spring day at his boarding school in Zuoz, Switzerland

Fig. 1.4 Tommy rowing on the lake near to his school

Schwarzköpfe (cream-filled cake glazed with Swiss chocolate), Aprikosentorten, and Erdbeertorten, accompanied by the beverage heissen Zitronensaft or Ovomaltine (Ovaltine). This is where our weekly allowances usually ended up. It also was the meeting center where one expected to find one's colleagues and where one could also get many other specialties of the Canton of Graubünden, like Bündnergerstensuppe (a thick barley and vegetable soup), if one was still hungry, or the favorite of the region "Bündnerfleisch," very thinly sliced dried beef, served on country bread with thick slabs of butter.

In summertime, there were other sports, like tennis and English-style cricket, for which the school owned large playing fields a little way below the village on the banks of the river Inn, which starts just a few miles up the valley from Zuoz and flows through Eastern Switzerland and then through Austria and into the Danube. We also had some competitive cross-country running through the natural terrain: forests or mountainsides, or even attempts to cross the river Inn without getting wet by jumping from one rock to another over the fast-running water. I enjoyed these exercises the most and became quite a good cross-country runner and also a fast uphill and downhill runner on mountainsides. All of these activities were useful in gaining me titles in track sports later in my college career at the University of Cambridge.

In Switzerland, my attitude to schoolwork changed completely, right from the first day. The classes were small, never exceeding 15 and mostly fewer than that. The teachers were really concerned that the boys followed what was being taught, and everyone was encouraged to ask questions if they had some uncertainty. I now paid attention in lessons and abandoned detective stories. Two hours every day were devoted to sports: a great variety was offered from which one could largely choose. Skiing was, of course, the great sport of the area, and there was excellent instruction. I enjoyed that the most and quickly became the best skier of my class; I think it was this that started me on a course of intense competitiveness, and soon I was just as keen to be the best in my class also in every subject of the schoolwork. I remember the intense emotion every time results of examinations were announced. There was just one competitor for the first place in my class, strangely enough in almost all subjects, a boy from the village of Zuoz, who just came for the daytime schoolwork. I was later told that the headmaster of the village school had recommended that he should be sent as a day pupil to our school, as he was too good to be taught adequately at his school.

I remember some particular episodes in the schoolwork that pleased me. We had a very nice, young biology teacher who was on good, friendly terms with the pupils. On the first day of the biology class, he explained what his demands would be: we had to keep a very neat notebook into which all drawings that he had presented should be copied at every lesson. He would examine these always at the beginning of the next lesson. These were mostly drawings of parts of plants and later anatomical details of animals. In the true Swiss fashion, drawing and calligraphy of the highest standards were expected. At the second lesson, he walked through to look at our copybooks. When he came to me, I had nothing to present. He seemed a little perplexed, and he wanted to give me a mild punishment for my disobedience. "Gold" he said, "go to the board and give us a summary of what we did the last time, together with drawings." I think he was expecting to trip me up, but he did not. I went through this, apparently to his complete satisfaction, and he said no more. The next lesson, the same little spiel was acted out. "What, you still have no notebook? Go to the board." This became the standard practice, and I believe he was quite pleased with it, since I obviously had to be particularly attentive to be sure I would not fail, and he could save himself the trouble of giving the quick recapitulation which he considered necessary. I got top grade in that course. I still find it better at any lecture to listen carefully than to keep writing in a notebook.

Another teacher whom I remember particularly well was the young man who taught us mathematics. Mathematics was a major topic there, and several branches of it were taught separately. This teacher was obviously a very interesting man, and I found the subject compelling. I thought of him, possibly quite correctly, as the most intelligent person I had come to know at that stage in my life. (Eventually, he became one of the leading figures in the Swiss insurance industry.) I think it is from him that I learned some subtleties of thinking, intellectual trickeries as one might call them, which, of course, are an important element in mathematics and which make mathematics a particularly useful subject for mental training.

I made many friends at the school, whereas I had made none in the 3 years at the Berlin school. The boys at the school came from many different countries—a major part from the Netherlands, Germany, and Great Britain but also from several other countries in distant parts of the world. One of my best friends became one Billy Benies, whose family had been the owners and directors of the largest German-speaking newspaper in Prague, the *Prager Tagblatt*. Billy and his younger brother Nikki Benies both turned out to be good skiers, and so many ski afternoons were spent together. These were the days of very wild skiing, not only in Switzerland, but in all other ski areas of the world, long before safety ski bindings had been invented and long before ski slopes were so crowded that everyone had to learn a certain amount of discipline. We schoolboys certainly felt none of that because we had no mechanical ski lifts, and every foot of downhill skiing had to be earned by doing the uphill climbing first. This was with "skins"—sealskins strapped to the bottom of the skis that made the uphill climbing easier and that had to be taken off at the top and wrapped around one's middle when skiing down again. In that situation, there were not very many people on any of the ski slopes, and therefore, we could practice the most daring exercises for which on a modern ski slope with a ski lift one would have one's lift ticket confiscated. We climbed high on the enormous ski slopes that we had at our disposal and then skied down even in very bad snow conditions in a straight line to do a carefully premeditated stop only where it was absolutely necessary. In some cases, this involved jumping over the roof of one of the huts that were placed around the mountainside to protect the cows that were kept on these slopes in the intermediate seasons. I remember talking later to some of the sun worshipers who were exposing themselves for tanning on the downhill side of the huts when a group of us mad skiers all came thundering down, one after another, over them, to land a little way below on the continuation of the ski slope. (Many huts were so constructed that a heavy snow blanket made the slope of the mountain continuous with the roof, so that a fast skier had good access to it). All this was great fun then, but such daring moves are now no longer possible with skiing having become so mechanized and the slopes having become so crowded.

Another friend I had at the school was one Maul Colledge, a brilliant athlete—tennis and ice hockey player—and the brother of the figure skating champion Cecilia Colledge who was the runner-up to Sonja Henie, previously the long-time world champion in that sport. Cecilia was a very good-looking girl, and of course, the displays of her skill on ice were a joy to watch. She would come to Zuoz most winters and spend the afternoons practicing on the skating rink of one of the good hotels in the area. I sometimes made my way there after skiing to meet Cecilia and her mother, who sat at the rink side knitting and supervising Cecilia's training exercises.

Many years later, after I had left the school and returned to my parents in Great Britain, I came to know Maul and Cecilia rather well. I was a frequent visitor at their house in downtown London, as Maul was at our house on the outskirts of London. Maul, at that time, had displayed his athletic abilities by becoming at the same time the center forward of the British ice hockey team and was also being seeded number seven in the Wimbledon lawn tennis championship. Maul was also in the business of flying a small plane, which his parents had bought him. He would fly me around, taking off from the famous Brooklands motorsport circuit and airfield, then down along the south coast of Great Britain, often on days with poor visibility. In the absence of any navigational equipment, Maul's technique was to fly low enough to read the street signposts to establish our position.

In the years 1933–1938, I had been enrolled in the regular Zuoz school program. During school vacations, I would travel to wherever my nomadic parents were at the time. These destinations included Vienna, Austrian resorts, sometimes even Berlin where we maintained an apartment as my father still had his business there, and we were still safe as citizens of Austria, then an independent country. Often I traveled to Italy where my father had good friends who owned a magnificent estate on the Italian Riviera near Genoa, and we were always invited to stay there. By 1937, my parents had settled in England, and this is where I went during school vacations after that date. I had to withdraw from the Zuoz school to my great regret, since I intended to go to Cambridge University. The hurdle was that no European matriculation examinations that I could take in Zuoz would qualify me for the entrance to Cambridge. Instead, I had to go to coaches who would prepare me for the Cambridge entrance examination, which included Latin as a compulsory subject. I had had not encountered this language in any school before. The coaches assured us that they could prepare me to get through that exam, learning no more than the minimum that their studies had shown to be sufficient for the purpose. In the winter of 1937–1938, we gave up on that coaching, as the Zuoz school offered to make a private teacher available who could handle the Latin, and I was allowed to return for one semester.

During our numerous vacations in Italy, my father had organized some industrial work that turned out quite successful. But then, Mussolini got tough with foreign investors and prohibited any export of money or goods from Italy, so that funds accumulated there were of no use to us while living outside Italy. That led to a very amusing episode for me and my sister.

A Fairy Tale

I was 17 and my sister 22, when my father made us the following offer: he said that there was a certain amount of money that he had, now bottled up in Italy. After a few months, it would be confiscated by the Italian government, and he did not like to support the work of Mussolini. He and my mother had spent some time there and lived quite extravagantly, but this had not made significant inroads on the capital. Would my sister and I assist? Would we go to Italy on a ski vacation and spend as much money as possible? There would be strict conditions: no attempt to take any money or valuables out. That would be too dangerous. Clothing and such items we could buy to our hearts' content, and, of course, we would stay in the best hotels and live a life of luxury, just so long as we did not do anything that would attract too much attention that would cause us trouble from the Mussolini government. This was all quite within the rules that applied at that time, only my father knew that these rules would soon be changed, as indeed they were.

The arrangements for our trip were that a company agent would meet us on arrival, bring up however much money we asked for, and pay whatever bills we had run up. However, we must not think that this was the lifestyle for which we were being brought up; it would just be a fairy-tale interlude. And, of course, that is what it turned out to be. It must be a rare experience to be asked to spend as much as is reasonably possible. I am sure even among the children of the ultrarich, this behavior is not encouraged. But we took on this task and made our plans.

We decided to travel first to Turin and stay there 3 days, just to buy up whatever personal effects we desired, and then go to Cervinia, in the Italian Alps, for 2 weeks of skiing. Then I would have to go back to my Zuoz school, and my sister would return to our parents in England. We traveled there with the absolute minimum of luggage, just enough to keep us going for a day. We had booked rooms at the best hotel, and we went out from there down the main fashionable shopping street. We had no idea what was what or how to negotiate with the various merchants. By the time my sister had bought three evening dresses and I had been measured for four suits, the entire district seemed to know us. Presumably, the information had leaked around the back doors of the establishments. At any rate, every shop we then entered greeted us at the front door, and the employees stood in line to serve us. The few other customers in the shop did not seem to matter. Of course, we had to have all articles prepared in a hurry. My suits, for example, would normally take 2 weeks to be made; I asked to have them in 2 days. "But, Sir, you must understand that we cannot work that fast, even if we do nothing else." But with a financial incentive, they could. I imagine that they went around the back door and arranged to share the work and the profit with their neighboring competitors, but, of course, we shall never know. We left after 3 days with all the goods. Having arrived with tiny handbags, we left the hotel with one old-fashioned cabin trunk and three large suitcases.

Then, in Cervinia, at the foot of the Matterhorn, we had a grand time skiing and partying. I was by then quite competent at skiing, and I relished what little fame this gave me. I would regularly ski down a little faster than the cabin lift so that I could get the maximum number of trips in that the lift could deliver. This involved taking a shortcut by jumping down over a wall, accessible from above but about 10 ft high below, that had been put there to deflect ski traffic from this route. But as I had mastered jumps of this height, I would take this route regularly. This led the owner of the local ski shop to ask me whether I would be willing to test a new brand of skis for him. I agreed to do this. In those days, all skis were made of wood, hickory wood for the best of them. So I took the pair of best, brand-new hickory skis up the mountain and went down my usual way, over the wall. But at the moment of impact on landing, something happened. I saw that I was now skiing with just half a ski on one foot. Evidently, it had split longitudinally from one end to the other, and the binding had held one half to my foot. I had to make my way down to the ski shop and advise the ski merchant not to buy that new line of skis. I had feared that he would be annoyed with me for having smashed up a new pair of skis, but far from that, he was delighted for the good advice I had given him.

Mussolini had been responsible, I am sure not intentionally, for this fairy-tale interlude. Shortly after we left, his government did indeed confiscate all foreign accounts, including my father's. A pity we could not have exerted ourselves more.

Chapter 2
Wartime Student Days

Admission to Cambridge University

My parents had emigrated from Austria to England while I was in the Swiss boarding school. However, when Hitler invaded Austria in 1938, I had to cut short my high school studies and return to my parents in England as quickly as possible, in fact the day after the "Anschluss" (the German annexation of Austria). The reason for the hurry was that my British Resident's permit might be revoked because my Austrian passport was now invalid. I was 18 then and would have had another year at school in the usual system. Instead, I flew back in a state of great excitement from Zürich to London on a DC3. For me, it was a dramatic situation: I might not be allowed into England where my parents were, and on top of that, it was the first commercial flight I had taken. I was so excited I fell off the steps of the DC3 in Croydon (London) where my father awaited me. (A good thing they didn't have 747s in those days!) Fortunately, I was allowed into England and soon applied to Trinity College, Cambridge, to study engineering or the Mechanical Sciences Tripos as they were pleased to call it. I had wanted to study physics, but my father felt unsure, despite my good school record, that I could make a living as a physicist. And he knew that I was good at practical things and so thought engineering was the right thing for me. I had to learn the specific subjects needed for the Cambridge entrance examination, like Latin, which I had not done at all, and English history. I did not have much trouble with the other subjects of the entrance examination. Mathematics, though different from what I had learned at school, I could learn easily, and mostly by myself.

Latin was another matter. I was quite good at learning it, but, of course, to make up for a 4- or 5-year high school Latin education in one year was not easy. With the Zuoz private lessons, I passed the Latin exam without really knowing any Latin. Another part of the exam was just fun for me. It was a foreign language, and as I was by then fluent in English, I made German my foreign language of choice. I remember walking out of the 3-h examination after 15 min, having completed it all, no doubt correctly. And I still remember the faces of all the other students shaking their heads and signaling to me they were sorry I had given up so quickly.

Thomas Gold *edited by* Simon Mitton, *Taking the Back off the Watch*,
Astrophysics and Space Science Library 381, DOI 10.1007/978-3-642-27588-3_2,

I was admitted to enter Trinity College, Cambridge, in September 1939, as a student in Mechanical Sciences. But on September 3, Britain and France declared war on Germany in response to Hitler's invasion of Poland. This was by no means unexpected by us: my father had told us already for some time that this was inevitable, despite Prime Minister Chamberlain's statement made almost a year earlier: "Peace in our Time." My father was really quite opposed to war; some might even have accused him of being an out and out pacifist. But then, of course, Hitler came along and eradicated pacifism the world over very quickly. But, if one needed a story to take one back to the pacifist sentiments of the World War I era, here is one he used to tell.

He had been drafted into the Austrian army and had become a lieutenant. One period of duty was on the Austrian front against the Italians, and he was an assistant in the army headquarters. Most of the time he would spend in a smoke-filled room, with a card table in the center. There was the general, the greatly admired Austrian commander of that front, with his senior staff. They were playing the Austrian card game of the times, Tarock, a point-scoring and trick-taking game. My father's duty was to bring the general messages from the telephone hanging on the wall. My father would pick up the phone and the message would be something like, "This is the commander of the so-and-so battalion at xyz. We are almost out of ammunition and now surrounded. Please advise us what to do." He would deliver this message in a whisper to the general. The great man would interrupt his card play almost irrespective of the point it had reached, put down his cigar, pull himself up, and bellow out at the top of his voice, "Halten bis zum letzten Mann" (Hold out to the last man). Then he would smile for a moment, basking in the admiration of his staff for being such an able and forceful commander, and he would quickly resume the interrupted game.

Despite such items as that in my education, and despite having gained admission to Cambridge, I applied to the Royal Air Force to become a fighter pilot. I thought this was the right thing to do in view of my intense hatred of the Nazis. My parents were upset because they suspected (correctly as it unfortunately turned out) that the life expectancy of fighter pilots was not good. I went through the medical and fitness examinations and various other tests and was declared suitable for acceptance. But then, the security services came out with the ruling that persons who had previously been Austrian citizens could not be accepted into the Royal Air Force. So I had to resign myself to entering the engineering course in Cambridge.

Those were not good times to start at the University. The courses were all shortened so as to make more manpower available for the war effort. The great departments such as the Cavendish physics laboratory were drained of talent. Much of the gentle lifestyle of Cambridge had been abandoned. Nevertheless, I came to like the undergraduate life. But so far as the teaching in engineering was concerned, I found this dull, and I much preferred to read exciting books in the sciences, such as Sir James Jeans' *The Mysterious Universe* and Sir Arthur Eddington's *Stars and Atoms* on astronomy, as well as books on embryology, books on biochemistry of living systems—all kinds of things quite unconnected with the studies I was supposed to pursue. As I found out later in life, these topics were the seeds of my future professional career.

Internment

On May 14, 1940, after just 7 months at the University, the disaster struck. The "phony war" was over, and the serious war had started. The Germans had begun the invasion of the Netherlands and Belgium and, almost immediately, France. The great "Maginot Line," the heavy defense system the French had built along the Rhine, proved utterly useless. It had been designed by generals who did not think that any German administration could be so evil as to march through the Low Countries (as they had done in World War I), where there were no defenses, and occupy the Maginot line from the back.

I was interned by the British authorities together with all persons with German or Austrian passports or identity papers in that area because in view of an expected invasion, the county of Cambridgeshire was judged to be in a strategic position, on account of its many RAF bomber airfield. People in similar circumstances in other areas, who had been cleared by the authorities as we had, remained free. My parents in London, as I found after some weeks of internment, were indeed free.

I was interned for 9 months, and although some of the experiences were dreadful and some quite frightening, I remember much of it almost with pleasure and certainly without hard feelings against the authorities, even when they subjected us, I think sometimes quite needlessly, to great danger and to serious health hazards. All these internees had undergone individual investigations and had been cleared of any suspicion of being pro-Nazi. We had been told that this meant we would be safe from internment in the event of war. But the police chief of Cambridge thought it best to intern us nevertheless. We were told it would be only for a few days, and we would be regarded as friendly aliens and given the treatment appropriate to that status. What followed was quite a different story.

The passports and identity papers of the Cambridge internees were all sent unintentionally on a ship to Australia, which in fact was sunk, but the internees, myself included, were sent on another ship to Canada. Probably it was due to some minor mistake on the part of an official. It was fortunate for me that he did not make the mistake the other way around. It meant, however, that when we were in Canada, there was no way of deciding who was who by tracing back to the previous investigations securely. Anyone could have claimed to be anyone else, and so this made the process of disentangling the situation much harder and no doubt our stay much longer.

When our ship was preparing to leave Liverpool for Canada, some of the sailors told us they had just heard the news that the last ship full of internees to leave the port of Liverpool, that one bound for Australia, had been sunk by a U-boat in the Irish Sea with all lives lost. Still, we made it to Canada but in pretty dreadful living circumstances for 2 weeks. We were a group of 800, crowded into the hold of a merchant ship, sleeping in three layers: on the floor, on the tables, and in hammocks strung above. Sanitary facilities now had to serve 800 persons, having been designed for about 20 sailors. Dysentery quickly became rampant. Very fortunately, none of the more serious diseases that are often promoted by such circumstances were present.

We were allowed out on deck only for short periods and then in shifts. The companion way to the deck had three single file fortifications spiked with barbed wire, evidently to defend against a mutiny at sea, a takeover of the ship by the internees. Of course, we all realized that it would also make any exit to the lifeboats extremely slow, if not impossible, and this was a consideration on everyone's mind.

Once in Canada, in a camp near the city of Quebec, things were better. We all recovered from the dysentery and slowly regained our original weight. The camp was pleasant with quite good facilities. Even the food was acceptable. The military commandant was friendly and knew that we should be treated as friendly aliens.

The good part of internment was the education. I feel sure that for me, at least, it was much superior to that which I was missing in Cambridge. Firstly, there was of course that hardening of personality that comes with facing such tough and serious conditions. If British education teaches one anything, it teaches one to try and believe that hardships are an essential component in the education of a young man. (Maybe so, but the pace-setting expensive British boarding schools have certainly been able to reduce greatly their heating and cooking bills as a result of this belief.)

But there was another side. It was that other scientists in the internment camp did their best to give science instruction. Hermann Bondi was the outstanding one so far as I was concerned. We had met on the first day of our internment, sleeping on the concrete floor of a disused ice cream factory near Cambridge, and with that beginning, we became friends for life. Like me, an Austrian from Vienna, he was the leading student in mathematics in Cambridge, and, as I found out later, greatly admired by the Cambridge mathematicians of the time. He taught me in easy conversation not only some details of mathematical techniques but also an overall attitude to reach for the important problems, to find quick and simple mental pathways to solve problems in physics, and to set one's aim at the highest level. He, with his brilliant undergraduate career behind him, found this natural. For me, at that stage, it was another matter.

There were a few other things I learned in internment that were more in the field of social studies. I saw how differently people behave when they are put in charge of a lot of other humans. For example, a representative elected by the internees on the ship that took us to Canada was beaten, on the instruction of the military commander in charge, when he explained that he was sent to discuss the bad circumstances, the dysentery in the big cargo holds of the ship. Again, in the second Canadian camp to which we were transferred, we had a problem with the commandant. He also knew perfectly well that we were all persons who had come to England to get away from Hitler. Nevertheless, he used every opportunity not only to be difficult but deliberately nasty. Some representatives from the camp were again beaten when they voiced complaints. I have a little story of an event that annoyed me particularly.

There were some 800 people in the camp. There was no shower and no warm water. Standing in front of a small wash basin with cold water and trying to wash down one's body was obviously rather unpleasant. Maybe some of the inmates avoided doing it, and that also was unpleasant. I had represented myself as a carpenter, and because I could read architectural plans, I was made carpenter foreman of a team of inmates who were to build new facilities for the camp (actually not for our benefit, but for use as an army camp at a later date). In the course of this work, I picked up some scraps of wood and galvanized iron sheet, and out of these, I constructed a small but adequate bathtub. The construction did not involve a single piece other than what was taken from the building scrap heap. The tub could be filled from a hose with cold water, and from another hose, steam from a locomotive boiler was available and could be used to heat it, indeed in a few seconds. The moment the bathtub was constructed, it was, of course, in the greatest demand. I immediately arranged for people to enter their names for a quarter of an hour each, all around the clock, so as to get the maximum benefit out of this overdue item of sanitary equipment. It was immediately booked up for several days ahead. This usage went on for a week, and then suddenly, I was called to see the commandant. I was a bit frightened, knowing that he previously had had people beaten. I assumed it was with respect to the tub, but I did not know.

When I was led into the commandant's office, he shouted at me, "So you are the one who made that bathtub!" I said meekly, "Yes Sir." "You were not sent into internment camp to recline in bathtubs, were you?" he said. I said, "No sir." He said, "I have given instructions to have the bathtub taken out immediately." Luckily, I was not beaten, but I was certainly very annoyed. The guards who took me back into the camp then picked up the bathtub and deposited it just exactly on the outside of the fence so that we could see it all the time. This commandant came to be so hated by the occupants of the camp that when the news broke that he had died (of stomach ulcers, in fact), the 800 members of the camp, I think to the last man, cheered loudly. Hatred was by then so deep. These circumstances seem, in ordinary life, almost incomprehensible. But having seen them makes clear that atrocities can be committed, such as those committed by the hordes of Hitler or of Stalin, by persons who would seem normal in other walks of life.

Beware of humans when they are put in charge of other humans. Natural selection, over the millennia of tribal human society, has no doubt encoded some elements of social behavior patterns into our genes. But will this cover the case when one man is put in command of the lives of hundreds, or thousands, or indeed millions of others? How can it? There could not possibly have been enough cases of that sort for natural selection to develop an appropriate response.

Back to Cambridge

The release from internment after 9 months happened when a benevolent man from the Home Office in London came over to sort things out. Of course, since we had no identity papers, he had to trust each person that he was who he said he was. He believed them when they explained what they had been doing, when they had come to England, and all of that. He drew up lists, which were really only priority lists of the sequence of release, but I think in the end, he had everybody released.

I was on the first ship to return, and it was quite a different matter from the trip going over. It was a Belgian passenger ship with cabins for two or three people, with toilet facilities, with a dining room (somewhat wave-battered by a previous storm, and partly boarded up), but all quite comfortable. However, the trip itself seemed very hazardous. As we were leaving Halifax, Nova Scotia, we saw a Tribal class destroyer of the Royal Canadian Navy getting ready to move out. The sailors on our ship told us that we would assemble with many other ships outside the port and travel across as a large HX convoy of about 50 ships, with the escort destroyer following us, but out of visible range. The reason for this was that at that time, one of the German pocket battleships had managed to get out into the Atlantic and had sunk other convoys, lock, stock, and barrel. Not a nice prospect, to be traveling out into the ocean as a bait!

Luckily, the German battleship either did not find us or it discovered that we were being shadowed by an escort. At any rate, we saw nothing of it. But we had another exciting and somewhat terrifying episode. The captain of our ship had told us to be on deck as much as possible, and on deck to spend our time searching over the ocean for U-boat periscopes. Of course, we all had a great deal of motivation to do this very diligently, and then one day, it happened. Between us and the next nearest ship, right in the middle of our convoy, we saw first the slicing through the water of a small object, and then the majestic rise of a submarine with the water flowing off first its conning tower and then its hull. Here it was in a few seconds—a large submarine.

Our ship had a small gun placed on its deck at the back, and the sailors manning this gun immediately started to move it. But before they could shoot, a flashing light signal came from the top of the conning tower. Fortunately, it was the identification signal that the submarine was one of ours. What a relief when this was announced over the loudspeakers! I feel sure that the submarine was there due to a navigational error. No submarine would want to surface in the middle of a zigzagging convoy. Presumably, the submarine had been following us in the hope of trapping the German battleship.

There was another skirmish, apparently with genuine U-boats, as we were approaching the Irish Sea. We saw the depth charges being dropped by British coastal command planes, but we were never told of the result. We arrived safely in Liverpool, and I traveled back to Cambridge in mid-January 1941, to resume my undergraduate studies. I had lost the major part of 1 year of the 3 year Mechanical Sciences course, which, due to wartime conditions, had already been shortened to 3 years from the original four. I was allowed to continue as if nothing had happened, and I managed to get through and obtain my degree.

Chapter 3
Wartime Work for the British Navy

Waiting for the Security Clearance

It was my association with Hermann Bondi that gave me the chance to enter radar research after graduating. Bondi had been hired by Fred Hoyle, the head of the theory division radar research establishment of the Admiralty, the British authority responsible for the command of the Royal Navy. Bondi had asked for me to be brought in as an engineer to join himself and Hoyle. Bondi suggested my name to Hoyle, just on the grounds that he knew me and he thought that I may be able to make a contribution to this most important and, at that time, highly sophisticated and top secret branch of the scientific civil service (Fig. 3.1).

It was quite ironic that a little more than a year earlier Bondi and I had been treated as dangerous aliens but we were now admitted to work in this most highly secret field! I knew that there would be a long waiting time, usually about 3 months, before the security clearances could be secured. I therefore had to find work for such a period, and I decided to seek it in England's Lake District which I knew well. I was accepted by the organization that produced the wooden pit props needed in the coal mines to hold up the roof of the seams from which coal had been taken. I had traveled to Greenock by bus, and I had in any case to register with the police whenever I traveled to a different location, being still under the registration requirements of Aliens Orders. So I went to the Greenock police station, and I told them of the lumberjack position for which I had applied and been accepted. The police immediately said that I would need to have somewhere to stay, and they said the only farm nearby was at Abbot Park. It was a 2-h walk to get there, but they would offer me a ride on a police motorcycle and take me there. They phoned Mr. Sharp, the owner of the Abbot Park Farm, who immediately accepted me, at least for an overnight stay. So I jumped with my backpack onto the police motorcycle and was taken for a hair-raising ride up the mountain and to a small farmhouse and farmyard which apparently was Abbot Park. Mr. and Mrs. Sharp immediately invited me in, gave me dinner, and showed me to my room, which was a very acceptable room though the general facilities of the house were very primitive.

Thomas Gold *edited by* Simon Mitton, *Taking the Back off the Watch*,
Astrophysics and Space Science Library 381, DOI 10.1007/978-3-642-27588-3_3,

Fig. 3.1 Thomas Gold, Hermann Bondi, and Fred Hoyle on the front row (as usual for Hoyle) at the XIth General Assembly of the International Astronomical Union, Berkeley 1961

I went to bed there and woke up in the morning with a fearful, very loud squeal as if someone was in extreme pain. I thought oh dear, what kind of a place have I landed in? What is going on in this house? I looked out of the window, and I had a view into the farmyard, and what was squealing was a pig standing in front of the entrance door of the house. Then I saw Mrs. Sharp come out with a bunch of keys in her hand and go to a shed on the other side of the yard and unlock it. The pig then nudged it open, having apparently been unable to do this when it was locked. It stopped squealing, went inside the shed, and came out with a large lump of coal in its mouth. It then settled down, happily chewing that coal, and the loud crackling noises made clear how hard it was. It ate up the whole lump while I was watching it and then retired elsewhere in the farmyard.

When I went down to breakfast, Mrs. Sharp apologized for the loud awakening that I had had and told me that the pig was in the habit of nudging open the shed door and taking coal from it to eat it. But coal was, as everything else, rationed in those days, and Mrs. Sharp was running short of coal for the domestic purposes for which it was used. So she had thought she could lock the shed and keep the pig out. But the pig, finding the shed locked, which was unusual, had started to squeal for what it thought was his normal right, to get his daily lump of coal. That was my introduction to Auntie Aggie. The pig had been given that name by the two sons of the Sharps, aged around four and six. They had called it Auntie Aggie, and we must assume that the resemblance to their real Auntie Aggie must have been striking for all members of the family to follow their example. When I learned more about Auntie Aggie, I came to understand the unusual importance that they attached to this little pig. While I don't claim to be an expert in the intelligence of pigs, I must say that I was absolutely struck by the intelligence and by the humor displayed by Auntie Aggie as I came to know her.

Over the first few days after my arrival, I was asked by the Sharps to help them on their farm rather than go to the lumber camp, and I was glad to do this. After a few days as a farm worker, I had to join the lumberjack work for which I had been accepted. I worked at that with great diligence but initially with not much skill compared with the professional. When this particular job came to an end and the whole operation was moved to another location, I was happy to stay with the Sharps who offered me the continuation of their farm work.

I had to drive a tractor out to the fields before the Sharps were up, and I observed Auntie Aggie watching me start from the farmyard and drive the tractor along a particular path through the woods. Then at one point where the path made a sharp turn, there was Auntie Aggie lying on the path in the middle of the way, obviously pretending to be asleep, but then how could she be when she had just a few minutes earlier watched me in the farmyard? Nevertheless, she jumped up as if startled by the tractor and cleared the way for me by running into the woods. I had seen her blink with an eye before my tractor had come close, and seeing that she was obviously trying to trick me, I laughed aloud when she fled back in the woods. This had obviously been a carefully staged event for which she had run through the woods to the point on the track that she knew I would inevitably come to, and then had positioned herself there trying to look asleep. Of course I told the Sharps about this occurrence, and they said wait, this is nothing when you come to learn what Auntie Aggie will do to produce humorous events. I certainly had to take note of the next event of that nature that I saw in the fields on one of the following days.

The two little boys were trying hard to ride on the pig who invariably shook them off, but was then very careful not to tread on them or hurt them as they fell. The boys had brought an inflated inner tube of a tire with them as a toy, and Aggie had started to play with it also. In the course of this, she got her snout into the tire and then dragged it until it got stuck around her middle, which, strangely, on pigs is the widest part. Then she could not get it off, because she could not get it to slip further back, which would have been her natural instinct. She ran along with the inner tube firmly fastened around her middle. But now as she ran, the front and the back hooves would bump into the tube and each time make a loud noise. We watched this and thought it was screamingly funny and roared with laughter at it, but this seemed to have given her an idea, because the next day in the same location, she would seek out the inner tube and slowly manage to get it behind her front legs and stuck around her middle. She then ran around the field several times with her hooves clunking against the tube, then seeing us again laughing about all this. So this time she was really not frightened by the situation as she had been the first time, but she had deliberately produced this as she had seen how successful it had been the first time. We then helped her to get it off, and she repeated the attempt several times more until she realized that we no longer regarded it as very funny and stopped laughing about it. Auntie Aggie was funny also in many other ways, for example, by picking up all the stray eggs that had been tucked away by the hens in the courtyard and for the most part when Aggie found them, they already had become very bad and were very smelly. The Sharps knew that bad eggs do not harm pigs, and so they were just glad to have them picked up. But Aggie would not just pick them up, but show us each one and then chew it open so as to display the smell to us.

Then sometime later, there was the terrible event when Aggie saw a brass ring that she had apparently recognized from before when it had been threaded through her nose for the purpose of preventing her from drilling up all the newly planted rows of potatoes and eating them. She saw that ring as Mrs. Sharp came into the pig's shed and immediately squealed and went to the far corner when she had understood the threat under which she had now come. Mrs. Sharp tried for a long time to approach Aggie, but she could neither hold her nor conduct this delicate operation. Aggie kept running back and forth in the shed and seemed to defeat Mrs. Sharp completely. Then quite abruptly she changed her attitude and presented herself for this painful operation which was performed with no more than a little cry from Aggie. She had evidently decided that she could not escape forever and she might as well have her owners have their say. She could then not dig up any more potatoes or at least not in the smart way in which she had gone along the furrows of the planted new potatoes before, but she got quite enough to eat in any case.

What had I learned in all this time in the Lake District? I had learned to wield an ax until I had huge blisters on my hands; I had learned to milk cows, to shoot rabbits (with a minimum of ammunition)—a major food source for us. There can be no question that Auntie Aggie remained the center of attention for me as she was for the Sharps. They had initially kept her as the one pig that each farm was allowed to keep to clean up its wastage, but Aggie had become the darling of the kids and of the parents also that the idea of sending her eventually to slaughter became unthinkable to them. When the time came for me to join the Admiralty radar establishment, I was really quite sad to leave.

The Admiralty Signal Establishment

I must say this for the British wartime scientific service. Under the pressure of this dreadful war, suddenly a lot of the problems that are normally embedded in the system of selection and promotion of scientists had disappeared. Seniority didn't count for much, pressure by peer groups did not matter, but good people were quickly propelled into important positions, almost irrespective of their principal subject, if they were seen to be clever, imaginative, and productive. Many toes of the long-established British scientific civil servants must have been bruised badly, but apparently some high-level administrators understood that this was what it would take to win the war.

It was quite an experience coming into the theory section at the Admiralty's radar establishment (radio location, as it was then called, at Witley, Surrey, in the south of England). There were six people in the section, all of them intelligent, likable, and witty. I felt I could not possibly have done better for myself than to join this group. The great enigma to start with was the director, Fred Hoyle. He seemed so strange: he seemed never to listen when people were talking to him, and his broad Yorkshire accent seemed quite out of place. I soon learned, however, that he had already made a very significant contribution in showing how the existing radar sets on British ships could be used to obtain, already at a 100-mile range, a reasonably good estimate of the altitude of an incoming plane. He did this without requiring any new equipment, just requiring the radar officer on the ship to plot the range of first appearance of the plane and the range of subsequent disappearances from the screen. What this was doing was identifying the position of the plane in the interference pattern between the direct wave of the radar set from the antenna to the plane, and the indirect wave, due to the reflection at the sea surface. A very simple method, but, of course, it was of the greatest value. Decades later (late 1980s), when the official history of British wartime naval radar research was published by the Naval Radar Trust, Bondi credited Hoyle with having made a major contribution to the effectiveness of radar during actions in the eastern Mediterranean.

Before long I also discovered that I had misinterpreted Hoyle's attitude of apparently not listening. In fact, he listened very carefully and had an extremely good memory, as I would find out later when he often remembered what I had said much better than I myself. I think he put on this air not to say, "I am not listening," but instead "don't try to influence me, I am going to make up my own mind." This rugged independence he maintained all his life, and I have certainly come to admire it greatly.

I learned a lot very quickly, with the constant help of my patient colleagues, especially, of course, Bondi. Soon I was engaged in useful work, finding out how to use the existing radar to best advantage against aircraft and submarines, in the presence of the disturbance of "wave clutter," the reflections of radio waves from a rough sea. I ran trials from a large radar trailer parked on the shore of Cornwall, later from a large instrument erected on the top of Snowdon (the highest mountain in Wales), and then from a station in Scotland, on the Firth of Forth. Mostly I was trialing by using the regular stream of Coastal Command planes that flew out and back for surveillance very frequently. Later, at the Scottish station, I was given the opportunity of organizing military planes according to my detailed needs, to optimize the different variables. It was amazing how easy it was to obtain cooperation, even when I did not have the authority to command it. Everyone seemed to understand readily that it was important what had to be done, and all the red tape that would normally get in the way simply wasn't there.

There were many branches of science with which I became acquainted in the course of this work. I began to understand radio antennas, and with this, the whole field of optics became much clearer to me. Later, when I taught optics, I would represent it essentially as a branch of radio science, just with a shorter wavelength. It still seems to me that this is the best way, especially now when optics is beginning to master many of the techniques developed much more easily for radio. Of course I also had to learn electronics, the electronics of the day being with vacuum tubes, quite different from the modern kind of integrated circuit or microchip. Nevertheless, the principles of receiver technology and of signal processing are basic, and that was a good thing to learn.

Bondi and I were living in a little house we had rented in the neighborhood of Witley, our base establishment. We were often joined there by Hoyle, and also by another friend who was to have a considerable influence on me, Richard Pumphrey, a biologist by training, but by then a crucially important member of the radar establishment. Discussion in the evenings was mostly about physics and astronomy, with Hoyle the driving force. "What could Hubble's observations mean?" (referring to the expanding universe) he would ask frequently. Bondi would sit cross-legged on the floor and work out some mathematical problems at Hoyle's request. I was mostly a listener, but I suppose it was in those days that I learned some astronomy, almost without realizing it. Those discussions also led eventually to some items of work in later years, which turned out to be important.

Tantallon

Tantallon (in East Lothian, Scotland) was the trials site the Admiralty had prepared for the performance testing of new radar sets, as they were installed on ships. The location was on the south side of the Firth of Forth, near a charming small resort town called North Berwick. After the latest radar set designed for the Navy had been installed there, I was appointed director of Tantallon. It would be my job to describe the changes in the range of first detection of incoming aircraft by the interference of the noisy signal of wave clutter. As I had been doing this type of work for quite a while by now, I was very happy with that appointment. My graphs would be distributed to all the newly equipped ships, and their radar officers would know the importance of a correct early warning of an approaching enemy aircraft. It turned out that this appointment had a number of extremely favorable attachments to it. I would be able to call upon naval aircraft stations in the vicinity to carry out flights for me and monitor their detection range in different sea conditions, but beyond that I would even be able to call on aircraft carriers in the neighborhood to send out flights just for me, in different sea conditions of my choosing, and so it was indeed an extremely good place in which to continue my work.

When I turned up in North Berwick, I found out about some other duties that this job entailed. North Berwick was regarded as a sensitive coastal area that might be subject to invasions by the Germans, and I would be the most senior government employee to decide whether any residence permission should be given to any person applying for a stay at North Berwick. I was not too happy about having to deal with the bureaucratic problems that this involved, but on the other hand, it had the advantage that the hotel treated me with the greatest care, since their trade was now very much dependent on my decisions. Of course I made the decisions strictly along the lines of whether any criminal or pro-German evidence existed against any person making such an application, but nevertheless, the hotel went greatly out of their way to please me in every respect that they could.

A Naval Air Force officer was assigned to keep in contact with me and provide me with aircraft to fly out from his air station, quite near Tantallon, and go out to the range that I had estimated would give me useful information. So most evenings I would spend with him in the North Berwick bar to discuss what flights I would desire and whether he could comply with my wishes. Commander North was his name, and he turned out to be quite a remarkable person. He was absolutely the person that Sir Bernard Shaw had in mind when he wrote the play Pygmalion (the musical and movie My Fair Lady is the screen version), where Professor Higgins was the person who could so magnificently decipher the origin of the speech of any person, and the area of their upbringing, and the area in which they had lived later in life. So after much conversation with me, he then decided one night, "Do you mind if I give you my analysis of your upbringing and of your early youth, as I would deduce it from your speech?"

In those years, I had not yet achieved the membership of many learned societies, which would cause my life history to be displayed in their membership lists, or through that in the many types of Who's Who. So at that time, he would have had no way of looking me up in any available literature, and his judgment must have been really genuine.

His analysis was "the first language you learned was in Vienna and so I assume you were born there. Then there is an episode of North German which could have been Berlin. Then you came to England where you have been until the present." So really Commander North, like the fictional Professor Higgins, had correctly deduced my background from my speech. But to Commander North I was only known as an Admiralty scientific officer in charge of a Navy research institute. North and I worked for a long time together, with very satisfactory results, and my Admiralty publication distributed to many ships of the Navy that depended in their critical decisions as to whether they might have to worry about an aircraft attack on the graphs that I had provided for them.

I had many other interesting experiences in that year at North Berwick, some pleasant and some extremely unpleasant. I had one very long night in fierce rain, returning to North Berwick from an appointment in Northern Scotland, and spent the whole night following an ambulance and assisting the ambulance driver many times when he got stuck in the water that had accumulated to a higher level than his engine could tolerate. He told me that he had a very seriously ill person on the ambulance and was trying very hard to get him as quickly as possible to the hospital in Edinburgh that would be ready to treat him. How we got to a nearby farmhouse in the pouring rain and cold wind, I cannot remember, but I remember arriving back at my own lodgings after succeeding in bringing him to the assigned hospital in the early morning hours, absolutely exhausted, wet, and cold.

But there were other more pleasant occasions, and some of them were the trips to the base point of Snowdon that I made in my little car for the purpose of studying wave clutter problems in the extreme conditions that occurred in the Irish Sea, which of course was seen at a steep downward angle from the 3,560-ft summit of Snowdon.

After a long and sometimes warm summer evening in Tantallon, I used to like to go down to the beach and go swimming in the Firth of Forth. I used to like to swim with what were then very modern items of swimming gear, namely, the flippers that increase enormously the advantage that you get from your legs both in speed and endurance and also the watertight goggles that allow you to see well under water.

Several times I had noticed when swimming a little way below the surface that there were quite a few lobsters crawling around on the bottom of the sea. I had often practiced underwater swimming with this equipment and learned to hold my breath for long enough to stay for as much as 3 min below. I thought about catching some lobsters, but they were big and looked rather aggressive.

One day a person on the beach told me that there were gadgets made for the purpose of picking up a lobster if you could reach it and then dragging it back to the shore. I went and obtained the gadget, and on my next trip out into the water, I dove down whenever I could see a lobster, and I soon learned to pick them up with this device. I occasionally gave one to the hotel in which I was staying, in North Berwick, and they cooked it for me. But then I thought how nice it would be if I could bring a batch of them back home with me on my next trip back to the south. So the day before such a trip was planned, I exerted myself particularly hard, and I managed to bring back four lobsters and put them in a cardboard box stored on the shore. That, of course, had taken four separate trips of exhausting swimming and diving, but all I could think about was the triumph it would be to bring four lobsters back to share with my colleagues. How would I handle the lobsters for an overnight trip? I obtained an ordinary sack made of the cheap sackcloth that was in common usage. I stuffed the four lobsters into it, wrapped in moist seaweed, and I hoped that this would keep them sufficiently moist. I tied the sack very firmly shut with a strong piece of string that I felt sure the lobsters could not cut. When I got on the train, I threw the sack together with my other luggage on the overhead rack.

For overnight trips on the train, it was usual to have all the lights shut off, except for one faint blue light that had to stay on. There were five other people with me in the compartment, all sitting up on hard benches, and all of them, like myself, trying to get some sleep. All of a sudden, there was a loud shriek from one of them, and another person jumped up and turned on the full ceiling lights in the cabin. There I saw what I had done. The person who had delivered the shriek was fighting with a lobster in her hands, trying hard not to be bitten, and the other three lobsters were crawling around on the benches or on other people's laps. When I looked at the sack, it was still lying on the luggage rack, but it had been slit wide open, presumably by the lobsters, who had then crawled out and fallen on the people sitting below. The train conductor was called, and I started to apologize for what had happened, and said that I would pay them compensation for their sufferings. Luckily no one was wounded, and I had offered them what seemed to me a very good compensation, but they were not satisfied. The conductor said that there was no question that we must get rid of the lobsters as soon as ever possible. Would they just throw them out of the window? No, the conductor said, that would be against the rules. But at the next stop, he would arrange to have someone pick them up. Luckily, at the next stop, nobody showed up with instructions to collect lobsters, and I continued with my trophy, freshly securely tied up. The lobsters survived the trip, although not too happily, because they had begun to dry up. But they still made a feast for me and my colleagues.

Excursions to Snowdon

Bondi, who was a keen mountaineer, decided to have the largest and most modern radar set installed on Snowdon in North Wales, from where we could overlook the entire Irish Sea. Bondi found a way of doing it with the help of only four unskilled laborers. When the diesel engines needed servicing, he would take them apart to fit new piston rings. When I visited him up there, which I did often, he was often drenched head to toe in diesel oil but very proud of his achievements.

The radar set on top of the mountain was there to see what a much smaller version radar set, but operating in a similar manner, would achieve when mounted in an aircraft. For calm days, the result was quite sensational. Any ship in almost the entire Irish Sea was clearly visible, and we could have detected a rowing boat leaving Dublin Harbor. But when the sea got rough, as the Irish Sea often does, things were quite different. Since I was by then the expert on the problem of sea clutter, the disturbing radar reflections from the waves, my visits there were to be at times of bad weather. One such occasion coincided with the worst storm recorded in the Irish Sea and on the Welsh mountains in more than 20 years. I had the good fortune of being on the top of the mountain before it started, to observe and take recordings of the damage which the reflections from the waves did to the radar performance. I worked all that night through the bad storm, but the next morning, while the gale was still blowing, I had to depart to go to a very important meeting that involved major decisions.

Getting down from the mountain was going to be a problem! The Snowdon mountain railway could not run in such a high wind for fear that it would be blown off the track and tipped down the mountainside in some of the exposed precipitous locations. I had to go down on foot. Visibility on top was just a few feet in the blowing snow. The wind was so strong that one would lean at 45° into it in order not to be blown over. I was to go down accompanied by three naval enlisted men who were working there under our command. But we had only gone about 300 yards when they asked me very shyly whether I would allow them to return to their quarters on top because, they said, mountaineering had not been part of their training. They were very shy about this because on no account could they refuse the command of a superior officer in difficult circumstances. But, of course, I let them go and proceeded by myself. At least there was the railway track to guide me, even in the worst visibility situations.

I soon found that the hard ice crystals were driving into my clothing and were scratching my face so that I saw blood on the gloves with which I was trying to protect it. Looking into the wind was almost impossible. The real difficulty came when I had to cross one of the narrow causeways that had been built up for the railway. It was hardly wider than the single narrow-gauge track, and very steep slopes were on both sides. Seeing how much I was buffeted by the wind while walking, I decided it was too risky to attempt to walk over this. What I had to do was to lie down, hold onto the rails, and just crawl on all fours over the dangerous sections.

When, after 3 h, I finally reached the bottom station of the railway, a warm fire and hot tea awaited me there, and I had to undress completely because all my clothing, both what I had worn and also what was in my rucksack, was completely soaked when at the lower levels the snow had turned to driving rain but still driving at a speed of 100 mph. It had not been a pleasant experience, but it certainly was a memorable one. The radar results I had obtained the previous night proved to be valuable information, although a gale of that strength was not anything for which one expected to be prepared. I suppose both sides in the war would have suspended operations.

A Spy Story

Many years later, we found out about a rather funny occurrence related to Bondi and myself and our radar set on the top of Snowdon. It appears that someone found a strange sheet of paper in an inn where I had stayed near the bottom of the mountain, and this paper was densely covered with handwritten block letters that did not make up any recognizable words. The person who found it handed it in to the local police, obviously thinking that it was some kind of coded secret message. The police handed it on, and eventually it went to the cryptology establishment who said that if it was a code, it was too sophisticated for them to break it. This, of course, then put the military intelligence on our tail.

We heard that a radio-receiving station had been placed at the bottom of Snowdon listening to the radar and any other radio signals that might come from the top. For the military intelligence officials, the great theory was that here were two persons of foreign origin—Austrian—and possibly they were the world's most clever spies. They had been able to get the British Admiralty to equip them with the best and most powerful radar set and to arrange to place this in a strategically magnificent position overlooking the Irish Sea, which was such a desirable hunting ground for the German U-boats. What, it was argued, if these two spies could signal by some simple code to the U-boats and give them the positions of every ship in the Irish Sea? Wouldn't that be the world's most magnificent spy story?

Well it seems that the listening post was there for some months but discovered nothing. The military intelligence communicated, of course, with the head of our radar establishment, who felt sure that this theory was nonsense. It was only many years later that he was allowed to tell it to us.

The Snorkel War

Another spy episode was not so funny. In the last year of the war, I had been promoted to the position of Deputy Head of a section called "New Devices." It was concerned with all kinds of radio trickery, jamming, antijamming, and new types of radar designs. One day I received a phone call from the Admiralty in London to say that a messenger was on the way to me with some very important document, which I should study immediately and without interruption until I could report back my findings. Two hours later, the messenger arrived, and he carried with him a stack of papers, blueprints of some kind, in a bundle that was about 2 ft × 2 ft × 4 in.

When it was unwrapped, I saw, to my astonishment, the stamp on each of the sheets "Deutsche Kriegsmarine" (German Navy), with a date stamped on that was only 1 week ago. Within a week of these blueprints being issued in Lorient in France, at a major U-boat base, some spy of ours had managed to steal them, get them transported across the channel to the Admiralty, who no doubt had their experts look at them, and then transport them to me for a further evaluation. Of course, I was never told how all this had been achieved. After the end of the war, we found out how sophisticated the British spy network was, especially in France where there were many supportive citizens.

What was on all these blueprints? I spread them out on the large floor of my laboratory. They were engineering blueprint sheets, but contained no description, no legends, and no explanations of any kind. The only writing on them were the dimensions. Of course, I couldn't complain that our spy had failed to pick up what it took to interpret the drawings. I spent the next few hours, all through the night, trying to see how the various drawings would fit together, assuming them to belong to one construction item. But what was it? Presumably it had something to do with U-boats, seeing that it had originated in Lorient. I puzzled out that there was something that looked like a float-operated valve closing a large opening. It eventually became clear to me that it was the design of a breathing tube to be mounted on top of the conning tower of submarines so that they could draw in air without exposing the target area of an entire conning tower. A small tube was all that would stick out of the water, and if a wave came that would swamp it, it would momentarily close. From the dimensions of the tube, I could judge that it could bring in enough air not only for the crew, but also to run the diesel engines.

I had in the past wondered often why such a thing had not been fitted to submarines. The usual practice had been, since World War I, to run the submarines at the surface with the diesel engines, but below the surface only on the very limited energy supply of batteries. Why not suck in air from the surface without exposing oneself? Well, I had been told that there were so many difficulties about that. Precise depth keeping was a problem. Wave swamping of the inlet when the diesel engines were sucking in air at a furious rate would cause a drastic pressure drop in the submarine that would be harmful to the crew and so on. Nevertheless, what I was seeing here was clearly such a device. I even saw that there were a number of variants of what I regarded as the attachment structure, so that obviously it was intended to be fitted to different classes of their submarines. Presumably it meant that the Germans were going to equip all their submarines with this device. And indeed, early in 1944, U-boats began to appear that were equipped with this device, later named the "snorkel."

The U-boat war had been an uphill struggle for Britain, and, of course, the amount of shipping lost to U-boats had been immense. Slowly we had gained the upper hand, largely because of the shortwave radar technology with which U-boats at night on the surface, charging their batteries, could be detected. What would be the situation now? I sent back a messenger to London with the papers and with my reply, stating my conclusions. I also made an estimate of the radar reflectivity of such a breathing tube, if that was all that would stick out of the water, and a conclusion that airborne radar would hardly ever be able to detect it, ship-borne radar only in a very calm sea. I heard from the Admiralty in London that they had drawn the same conclusion in the first place and wanted me only to confirm it. Clearly a new phase of the U-boat war was going to begin: the phase of the snorkel.

The phase of the wholesale destruction of U-boats on their way out from the French coast by airborne radar was going to come to an end. The gains in the U-boat war we had made in 1943 might be reversed by this invention. What else could we do to counter this? A successful invasion of Europe and a defeat on land of the German army was the great hope of everybody who knew what the situation was. But until then, the only possibility lay in the heavy bombing campaign of all the harbors that were used for U-boats, in the hope of destroying either the U-boats themselves or the various facilities that were required for their servicing. That indeed was done, and insofar as we could find out, it was quite successful.

I was commissioned to find out experimentally how well the snorkel could be detected by shortwave radar in different conditions of the sea surface and from antennas mounted at different heights on ships. Soon I constructed some buoys that could be anchored and whose overwater exposure was that of the snorkel. One of the urgent questions was at what height should a short wavelength radar be mounted on a ship for best performance against a snorkel. I found a site on the South Coast of England, at a headland called Beachy Head on the channel coast, where a radar trailer could be towed to different levels above the sea. A 12-ton trailer was equipped with a navy radar set that would be used on ships. Ten snorkel buoys were fashioned and anchored in the sea, distributed from nearby out to ten miles away. A motor torpedo boat flotilla was stationed at nearby Newport, and they would do the somewhat risky task of taking these objects out and anchoring them just a few miles from the German-held coast.

I was given a truck that was normally intended to pull a 40-ton tank on a trailer, and I was to tow my radar trailer to the site. I don't think I had ever driven anything larger than the little family car. It was quite an experience to drive suddenly about the largest equipment that I had ever seen on a British road, maneuvering it through small villages where truck and trailer would only just fit through the narrow streets, and then to drive it up the grass slopes on Beachy Head and quite close to the crumbly vertical cliff edge. But this was the quickest way of testing out the radar set from different heights, since I could just drive my trailer up and down from a height of a few feet above sea level to about 150 ft.

It all worked out, and within 3 weeks, I had all the data of the range at which the snorkel could be detected in different sea conditions and with different heights of the radar antenna. In a calm sea, the performance was quite good, and I could see the furthest out buoy most of the time. One morning it wasn't there, and I looked around with my radar and found a similar target but two miles from where it was supposed to be. I got a motor torpedo boat to go out, risking that it might encounter the German boats, but I could assure them that I would detect a German boat very much further out. Our motor torpedo boat went first to the site where the buoy was supposed to be and confirmed that indeed it was not there. I then directed it to where I thought it was, and they still had difficulty spotting it, even when they were within 50 yards of it. They were most surprised at the precision with which I could direct them, and the motor torpedo boat flotilla was quite impressed with the radar performance.

I came to know much of the motor torpedo boat lore of the day while I was working there. The German MTBs, bigger, faster, and diesel powered, were clearly a much superior craft to ours, both for the armaments carried and the smaller vulnerability that came from the use of diesel fuel rather than the much more flammable gasoline (petrol). Nevertheless, most nighttime skirmishes between the MTBs had gone in favor of the British, and the superior radar was the main reason.

The Germans had a radar set on their motor torpedo boats also, but they were instructed to turn it off because the disadvantage of giving away their position to radio direction finding outweighed any advantage that their poorly designed radar could give them. In fact, it was said that the "minimum range" of their radar set, corresponding to the time delay of the recovery of the receiver from the emitted pulse, was so great that it exceeded the range at which a British MTB could be detected. They could do nothing with it. After the war, I had occasion to inspect these sets, and indeed that seemed to be true. I supposed they could have detected a battleship, but that was not a very likely contingency.

After the trials at Seaford, I also flew in military planes with airborne radar trying to detect the snorkel buoys. The results were dismal. I remember how the airplane crews that flew me understood so clearly how bad this was, and how, so late in the war that looked as if it was being won, a new and very serious threat for Britain had unexpectedly appeared.

My trials at Seaford and the airplane trials all predicted a rather gloomy future for our war against the U-boats. The methods of the past that had been moderately successful were all now outdated. Only two things remained to prevent a disastrous blockade of Britain. One was the bombing of the U-boat shipyards and harbors, and the other was the hope of a successful invasion across the channel, the "Second Front" Project, as it was called. Luckily only a few of the new German U-boat fleets, designed for snorkels, faster and more powerful than the old refitted ones, got into service in time before the invasion at the Normandy beaches on June 6, 1944. The starving out and defeat of Britain by the snorkel fleet did not happen, although the German navy had been very confident of victory almost up to the end. I was to hear of that later in my interrogations of numerous U-boat captains after the armistice. Naturally we went on working as best we could, and the British High Command no doubt knew when the invasion of France was to take place on which everything would depend.

Another radar-related important episode in the Naval warfare is worth relating, although personally I had very little to do with it. One of the British admirals had come to visit our establishment, and instead of being enthusiastic about all the marvelous new radar sets we were creating and the superiority we had in that field over the Germans, he was quite derogatory. "What we need on our ships are guns, not Christmas trees." Christmas trees referred to the complicated fanciful looking structures that filled up the masts of the ships, namely, all the radio and radar antennas. I wondered at the time why he had bothered to come to us if he was only going to be discouraging.

Shortly after this, it so happened that one of the fast and greatly feared German battle cruisers, the *Scharnhorst*, had left a northern Norwegian fjord, where it had been poised, clearly in order to intercept a munitions convoy of ships returning from the run to the Soviet port of Murmansk. British navy ships were protecting the convoy and had fought off the *Scharnhorst* twice, following which it turned and ran. I presume this was the success of the "Christmas trees" on our ships that could see better than the *Scharnhorst* in the stormy arctic night of Christmas day 1943.

The *Scharnhorst* had suffered some damage in the first encounter by a hit from a British battle cruiser HMS *Belfast*, but it still had superior speed to all except the British destroyers. They continued to threaten it, and meanwhile the powerful British battleship, HMS *Duke of York*, had joined the fray. *Scharnhorst* lost speed, presumably due to the damage, and the *Duke of York*, still unbeknown to the German ship, together with four destroyers, could now follow it and prepare a battle plan. The British ships more or less surrounded it, and then, suddenly, a starshell was fired up over the *Scharnhorst* illuminating it brightly and making it instantly a target for the optically controlled heavy guns and torpedoes of all the ships.

The admiral in charge of this carefully planned maneuver, Admiral Fraser, was sitting, so it is reported, in front of the plan display radar tube of the *Duke of York*, and he watched in marvelous detail what was unfolding. But soon the blip in the middle of all the ships, that represented the *Scharnhorst*, just disappeared. The admiral shouted with rage, "Just like the goddamn radio-location (radar) to pack up when the going gets hot!" In reality, of course, the going wasn't getting hot at all—it had cooled off completely, because the *Scharnhorst* had been torpedoed by the destroyers HMS *Scorpion* and the Norwegian *Stord*! I suppose that after this Battle of North Cape, the admirals revised their opinions about our "Christmas trees."

Ultrasound

One of the last major lines of scientific research that I undertook in the Admiralty toward the end of the war was technically perhaps the most interesting. It was the construction of a "moving target indicator," a radar system that could sharply distinguish between stationary and moving targets. It could also have been adapted to distinguish between radar reflections from the waves and the snorkel. Luckily, the war had come to an end before it could be so used.

For this system, I had to develop a number of quite new components, among them the memory units required to compare one repetition cycle of the radar with the next, so as to note what change on the small scale of the radio wavelength (10 cm) had taken place in the short time interval of the repetition cycle—in this case 1/500 of a second. I decided to translate the returned signals into sound waves and use the "slow" speed of sound to make the previous signals available for comparison with those of the next repetition cycle. These delay lines, as we called them, used mercury as the transmitting medium and sound at frequencies far above the audible range—in fact up to 13 MHz—resulting in a wavelength as short as 1/10 mm. When I had perfected this device, I could obtain accurate replicas of the signals, delayed by 1/500th of a second. I found the experimentation with ultrasound most interesting in itself, and I noted many effects that had not been known. The field was really quite unexplored. For the purpose of studying the effects of ultrasound in liquids, I had made myself a mini-radar system using this ultrasound, and I could "look" at many objects with this. It was just like the real radio-wave radar but scaled down by a factor of about 200,000, the ratio of the speed of radio waves and light, to that of sound in water or mercury.

One day I turned this radar on my hand. The result was startling. I immediately got a clear image of bones and tendons in my hand. Immediately this reminded me of the description of the discovery of X-rays by Röntgen some 50 years earlier. The value of this technique for medical applications was clear. It would be a second method for looking into the interior of the human body but with different detailed responses and with no harmful side effects. I could also take pieces of steel and detect cracks in them, but in this case very much better than with X-rays. X-rays can find a crack in steel only if the orientation is just right, so that the X-rays pass through in a direction of the crack and encounter less metal. In any other direction, the effect of the crack is undetectable. For my high-frequency sound wave, it was quite otherwise. The crack was an enormous barrier and showed up in virtually all directions in which the piece of steel was presented to the sound. Testing steel welds for cracks was, as I knew, a very important matter in technology, and this, it seemed to me, would be the best method. I was fascinated with these discoveries, but at this stage, I was still employed by the Admiralty to do radar work and not think of anything else. Later, after the end of the war when I left the Admiralty service, I tried to further this subject for its applications to medicine and to material testing, and I will return to this point later.

I suppose it was information about my work with such devices that brought me one day a very interesting visitor. His name was Alan Turing, a mathematician who came to discuss with me the possible technology, including the memory devices, of the great computers of the future, which he foresaw. I did not know, and he was not allowed to tell me that he had overseen at Bletchley Park the construction of the Colossus computer with which he had been able to break the German secrecy codes used in virtually all their radio communication systems, including the Enigma cipher machines. It was this that had given Britain and the allies a great advantage in all tactical situations of warfare.

But still I knew nothing about this when Turing came to visit me. What impressed me was firstly how strange a person he was, extremely shy and modest, and yet apparently immensely knowledgeable and imaginative. He discussed with me how electronic computers would take over not only mathematical computations but most of the commercial administrative work. Payrolls, bank accounts, all this he realized already then, in 1944, would become the prerogative of the computer. I don't think there were many people at the time who foresaw all this so well. I saw him two or three times more. He died in 1954 at age 42 in tragic circumstances. He and von Neumann at Princeton are now regarded as the pathfinders for the evolution of computers.

Preparations for the Invasion of Europe

The invasion of Europe was being prepared early in 1944, as we all knew, but the precise date and location were at the highest level of secrecy and not known to us. Nevertheless, I was asked to contribute some radar expertise so that each ship would find securely the precise landing spot on the coast to which it had been assigned. Hundreds of ships would be involved, and it had to be a precise system to keep them out of each other's way and to land in the right places. They all had radar sets, and it was a question of devising for each ship a map that would show precisely what the coastline would look like on its radar during the approach, so that each ship would make the corrections to its course. Transparencies had to be produced for each ship that could overlie the plan position indicator radar display, so that the navigator could see whether the ship was on the correct course or should move a little to the right or a little to the left. All these radar maps had to be fashioned just from the cartographical material available for the coastline, so we had to predict what would show up on the radar just from an examination of maps with, of course, their contour heights marked. We could not take any actual radar sets into these locations, and of course there would be many unknowns about the radar reflectivities, the detailed roughness, and the electrical properties of many parts of the shoreline.

Nevertheless, I remember deciding that the essential thing was just to plot all the land areas that would be within a straight line of sight from the position of the radar antenna and its height above the sea surface. Detailed contour maps of the land could provide the information with which to draw such overlays for the radar set. The different reflectivities of different areas of land need not concern us, I argued, because we could easily turn up the gain (that is the power level) of the radar sets so that they would see every area of land that is within the line of sight, but they would still see nothing outside of such areas. The Earth's curvature did, however, need to be taken into account. The overlays we would prepare would closely match the picture on the screen, and one could slide them to the right or left for best match and read off the error in the ship's position.

I set about drawing such overlays laboriously by hand from maps of some arbitrarily chosen section of the French coastline, choosing a point a certain distance offshore, the height of the radar antenna as it was on most of the landing craft; I then drew lines radiating out from that point with a routine that I would mark the line only where no forefront object would have obscured and where the contour heights of the map were correctly recalculated, as they would have been drawn by a cartographer who had assumed the Earth was flat and had taken all heights above or below the tangent plane from the assumed position of the ship.

With enough such lines plotted radiating out from a point, one could of course then delineate the areas that the radar would see. It would have to be done thousands of times over for the many ships involved and for each ship for various distances from the shoreline. It would be an enormous amount of work. To make it easier, I designed a small cartographical instrument where one drew each radial line with an automatic adjustment for Earth curvature built in by a parabolic guide on the instrument, where one had a setting for the height of the radar antenna above sea level, and where one would then merely draw the line whenever the contour heights exceeded a certain figure you could read off on the instrument. With this device I could do the job at least ten times faster than before. I submitted this, together with all my considerations on the subject, to the headquarters organization who had asked me to look into the problem. I hope they used it, but whether they did or did not was not for me to know. I know the Admiralty took out a patent on my little gadget.

In the course of the preparations for the invasion of France, I also traveled up to Inverness in the north of Scotland, a night and half a day on the railways as they then were, to join landing trials with tank landing craft practicing on the southern shoreline of the Morey Firth. I arrived there, got on one of the tank landing craft, and started to operate its radar set as we left Inverness harbor. It was to take some 6 h to reach the designated landing spot on the Scottish shoreline in very rough, stormy weather. Landing craft are designed to ride up on a beach, but they are not designed to make their way through big waves. Although I had normally been quite resistant to seasickness, this was more than I could take. I operated my radar set until I was so ill that I could no longer see what was on the screen. I collapsed on a bed, and I lay there throughout the entire landing maneuver and throughout the entire trip back, and I recovered only when the ship reentered Inverness. I remember only the advice that the Scottish captain gave me, and that was, "You should not eat anything other than strawberry jam on occasions like this." He was waiting for my question "Why?" "Because it will taste the same on the way up as it did on the way down!"

I got on the next train back, having achieved precisely nothing. It is a good thing that when D-Day happened, the waves in the channel were not as high. Otherwise, I don't think many soldiers would have been able to do as much as get off their craft, let alone fight the enemy. The only thing that I had learned on the trip was the robustness of tank landing craft that would smash down from a height of perhaps 15 ft, with their flat belly against the sea in a tremendous crash, and survive even hundreds of impacts of that kind.

There were some other radar-related activities with which I had come into contact, and though I was not directly responsible for them, I had perhaps contributed to the discussions. One was the deliberate painting of false targets on the enemy's radar set. Churchill was a master of wartime deception, and his advice in every area of warfare had filtered down into a local decision-making process.

Churchill's statement to his military commanders had been "Truth in war is a very precious thing and it must always be accompanied by a bodyguard of lies." Well, of course, radar gives away to the enemy one item of undoubted truth, namely, the powerful radar pulse which can be seen at enormously greater distance than that from which radar echoes can be usefully received. The enemy ship gets the truth about the direction to our ship, though not the range. Still, how to accompany this by a bodyguard of lies?

Well, what you do is you plot the rotation rate of the enemy's radar set, which is usually a perfectly fixed quantity for a given set. You will then know when it is facing a direction away from you. In most cases, you can still detect the transmitted pulse even then, because there will always be some scattering in the vicinity of the transmitter that will put a little power back toward you. You can then arrange to radiate a pulse or pulses on the same frequency with certain time delays, and he will receive these again even from the back of his radar antenna. But of course he does not know that they came from the back, and he will assume that they came from the front as normal radar reflections. In this way, one could paint on his radar set what would look like a whole fleet of ships advancing from the other side toward him. You may then predict what his response to that would be, and gain a great tactical advantage. I don't know how often these sophisticated tricks were tried, but I heard later that they were quite successful on some occasions.

Many other "bodyguards of lies" accompanied other radio techniques. The German directional radio beacons with which the nighttime air raids on Britain were guided were often overpowered by false radio beacons which misdirected entire bomber fleets to irrelevant locations. A few German bomber pilots were so misled by the false readings on their navigational instruments that they even flew as far as Ireland. I am sure this was not a deliberate part of the British misleading techniques but merely an involuntary consequence. How much of the victory of World War II we owe to Churchill, not only as the principal strategist but also as the viciously cunning commander of all types of operations, we shall never know. From what I came to know about it, it seems much more than is publicly known.

My father, who was a committed anglophile, was also a great admirer of Churchill. He would remember almost verbatim many of Churchill's speeches, spoken as with the Elgin marbles in his mouth like Pericles in ancient Athens. If one has lived through these somber days, with its hardships and its perils, and with the thought of a Nazi victory being one of indescribable horror, then one remembers the importance to public morale of these speeches. But beyond that, Churchill's wit and repartee were a constant source of delight. "A sheep in sheep's clothing," as he said of Attlee, the very mild leader of the opposition labor party. "There but for the grace of God goes God," he said of Sir Stafford Cripps, the very pious and self-righteous senior labor member. Or, when a new member of parliament delivered his maiden speech, and Churchill sitting in the front row opposite, darkened his brow and shook his head, which so unnerved the new MP that he interrupted his speech, looked at Churchill and said, "but I am only expressing my own opinions." Churchill was heard to murmur "And I am only shaking my own head."

These and many other stories just made Churchill the great hero to so many of the British public. It is not often that a politician is also a superb military strategist as well as being a first-class writer, a wit who could dispense sharp criticism, and even a moderately good painter. Churchill was certainly better than Hitler, who also indulged in painting.

I had the opportunity to meet him personally after the end of the war, when, as a Fellow of Trinity College in Cambridge, the Master of the College, George Trevelyan, was showing Churchill around the college and cornered me, a young Fellow, to be introduced to Churchill and chat with him for a brief moment. He appeared to be just a jolly elderly gentleman, full of good humor and a face that gave expression to every word he said. As an actor he would have been in the category of a Charles Laughton.

But we will return to the grim realities of the time. As the preparations for the invasion of France were proceeding, the French channel coast was kept under almost constant bombardment by our airplanes. One such striking force was a Canadian contingent who flew these bombing missions early every morning, mostly with chemically timed bombs that could not be disarmed in any way. An acid inside was just going to eat its way through a diaphragm, and when it did, the bomb would explode. Nothing you could do from the outside would stop it; the most sophisticated bomb disposal squad could do nothing with it, even if they knew all the details of its design.

The trouble for us was that this Canadian contingent was operating from an airfield adjacent to the house in which Bondi and I, and Hoyle some of the time, were living. In fact, it was our house that was the first object the heavily laden planes had to clear on takeoff. When we had rented the house, we did not know of this particular drawback, but now we were stuck with it. After a while of being awakened by 20 planes in succession just clearing the rooftop at 4:30 a.m., we got quite used to this and could sleep through it.

But then one morning I woke up in a state of shock—there had evidently been nearby a violent explosion. I must have been sleeping with my mouth wide open, for a large chunk of the plaster from the ceiling had fallen into it. As I was spitting to get it out, my bedroom door opened, and Fred Hoyle, who was staying there at the time, stuck his head in and said "Did you hear that?" I said, "what do you mean, did I hear that? The house nearly collapsed!" He said, "I know, but I heard, about 20 min ago all the planes taking off except for one, where I heard the take off noise just suddenly stop, and then nothing more." "So," he said, "I went back to sleep, and then came this noise, which of course, woke me up." I said to him, "How can you be so stupid, to go back to sleep, when clearly what must have happened was that the plane failed to take off, caught fire and its bombs exploded?" He said, "Well, I know that now, but I couldn't have done anything about it anyway."

We later learned that this is exactly what happened. The crew had been able to save themselves, but the burning wreck eventually exploded its bomb load. It was only about a hundred yards from our house. Hoyle's attitude, to go back to sleep after hearing the obviously failed takeoff, may seem incomprehensible to those who did not know him. I am sure I would have rushed out of the house as fast as I could in the direction away from the airport. Hoyle was more relaxed about it.

This was by no means the only explosion near our little house. It happened several times that returning Canadian planes had been shot up over the French coast so that their hydraulic bomb door mechanism failed. They then had this chemically fused bomb that they could not drop, and all that they could do was to return to the airport as quickly as possible and have the doors opened and the bomb removed and put into a pit to await its explosion. This pit also was quite close to our house, so it happened that on many days, when we came home from work, we would find some bomb damage, like broken windows and a crack running right down the middle of the front wall of the house having widened by another inch. After a few events of this kind, the local glass merchant, who apparently was paid by the government to repair all bomb-damaged glass, seemed to have come out after hearing the usual bomb explosion, and he would replace all the usual broken glass in our house, so that when we returned, we merely found fresh wet putty in the windows and dirty new glass panels in place. There was no other information, no paperwork, nothing. The glazier just knew his job.

There was another bomb incident that is worth recounting. One evening our friend Pumphrey was staying with us, and after dinner we had settled down to some pleasant conversation in the living room, and Pumphrey said "How wonderfully quiet it is here. Where I live there is a tank training school, and all night long we can't sleep because of the rumble of large numbers of tanks all around our house." I am not exaggerating, but just then, when he had finished talking, there was the noise that we all knew too well, that of a high pitched whistle getting louder and louder—the noise of a falling bomb! We could hear the engine noise of a solitary airplane high in the sky. Of course, we all dove to the floor, under furniture, in the way we had learned. And then there was the WHAM of the bomb. The front door of our house was blown in, and some further damage was done to this battered house. We rushed out into the dark night to try and find the location where the bomb had actually hit. A field quite close to our house was the place, and we found the crater, and still-hot steel fragments of the bomb in it. I am not sure that Pumphrey did not prefer the rumble of the tanks all night to this particular visitation!

With our interest in astronomy, and my work with large powerful radar sets, it was not very farfetched to think of the application of radar to the study of the Moon and the planets. I made a little calculation sometime in 1944 to determine the power and antenna size that would be required to get an echo back from the Moon, or from Venus or Mars. The Moon, I decided, would be quite easy. I wrote a paper for the benefit of Admiralty headquarters, entitled "Radio Communication via the Moon." What I suggested was that a secret radio channel could be established between any two locations on the Earth that could both see the Moon at the same time. This would enable the Admiralty in London to make contact with their ships at different times over very wide parts of the globe. It would be a communication channel far superior to the shortwave communication by reflection from the ionosphere, and if it was done at a high radio frequency, an enemy reception of the signal was almost impossible. I did not realize at the time how good the Moon's surface was for this purpose. I thought that the different time delays for reflections from different parts of the Moon would scramble a voice channel, but perhaps it would still allow something like a Morse channel. What we found out later was that the Moon's surface is so smooth that almost all the power is returned from a small area in the middle of the disk and that a voice channel, though not perfect, is quite intelligible.

Again I do not know how far this suggestion went at headquarters, but many years later, very serious attempts were made in the USA to eavesdrop on Soviet internal communications by directing a large receiving antenna at the Moon and picking up the small amount of radio power that had come from the Soviet Union by this route. It was not a success. I suppose because the antenna would pick up simultaneously a large number of scrambled signals from different regions, when in their intended use, the Earth curvature was preventing such interference.

Radar to the other planets, I realized, would require a very large antenna. In terms of the transmitter powers and receiver quality we had, I could calculate how large and decided that it would have to be done by some antenna of such size that the main reflector would have to be immovable, firmly fixed in the ground. Twenty years later, I had such an antenna built under my auspices in Puerto Rico, and indeed capable of obtaining echoes not only from Venus or Mars but soon also from Mercury and from the satellites of Jupiter. All that I had thought about then came true. We could measure surface properties, we could define distances and orbits to a vastly higher precision than before, and the surveying of the inner solar system was on a completely different footing.

After the successful Normandy landings, the war must have looked totally hopeless for the Germans. It was only the madness of Hitler that would prolong the fighting for another year, achieving nothing other than the slaughter of a vast number of people on both sides. The drama had to be played out to the final bitter end, like in a Greek tragedy.

The Germans' Surrender

When at last the surrender came, on May 8, 1945, the German navy was given instructions where to go and hand over their ships. The allies remembered, of course, what had happened at the end of World War I, when the German navy was supposed to surrender in the large harbor of Scapa Flow, and instead of handing over their ships, they sank them all. The victors hoped to avoid a recurrence of such a "surrender."

To that end, all the German U-boats were directed to go to the northern Irish port of Londonderry and tie up there. A small number of experts, myself included, were to fly out there to inspect and record any technological innovations that they might have on board. My task was to look after all the electronic equipment. When I arrived at the port in Londonderry, I suppose 3 or 4 days after the date of the surrender, the harbor was choked full of U-boats. I don't think anybody had ever seen so many U-boats in the same place. The number, as I recall, was 49, and I had to inspect them all. I was paid double my normal salary for the time involved as "danger money," because there was the fear that they would scuttle the ships. But nothing of that sort happened. I remember climbing down and up and down and up, conning tower after conning tower, knocking my head on innumerable bulkheads, and being nauseated by the submarine smell. Many of the U-boats had been at snorkel depth for a matter of weeks before hearing the instruction to surrender, and the ventilation that the snorkel provided was evidently very poor. The electronic technology that I found contained nothing new, but perhaps I noticed in general the more solid construction of the components and the very good quality of workmanship. There was not much I could report about that.

Another aspect of my inspection was much more interesting, and that was my discussions with many of the U-boat captains and senior officers. Obviously I conducted interrogations in German, which was a great help because not too many of the personnel had learned much English. They were mostly very young men; even the captains gave the impression of being almost fresh out of high school. The U-boat losses had been enormous, and so not much time had been devoted to training new crews. A few had survived one or even several sinkings of their ship and described the circumstances graphically. The most common was "we came to the surface at night, out from a French port, and before we knew anything else there was a searchlight directed at us from the sky and seconds later, before we could dive, came the explosions." This represented the success of the anti-U-boat patrols by the British radar-equipped planes, which could search for U-boats over large areas of ocean and, having detected one, could then fly quite close before their presence was known, and switch on their searchlight only during the final bombing run.

I discussed with some of the commanders the inferiority of the German military effort in the field of radar. They all seemed to know this perfectly well, and one of them said the German high command didn't take any notice any of these clever things. If it wasn't something that could make a big explosion, they were just not interested. Another U-boat commander was quite defiant. He would only talk to me in the brief terms that he had been instructed to use if he were a prisoner of war: name, unit, and serial number. But then he allowed himself a remark which was, "Well, we had to give in this time because the army was mismanaged. You know full well that we would have killed you at sea if we had been able to continue. It won't be long before we will fight the next round."

All those submarines were then taken out to sea by British navy personnel to be abandoned there and blown up. I did not understand why, but I suppose the argument was "the only good U-boat is a sunk U-boat."

Travels as Wing Commander in the Royal Air Force

Early in 1946, I was summoned to London to the Air Force headquarters, to be given another assignment. When I arrived there, I was told that I had been selected to head a small delegation to travel around Germany and visit many scientific and technological centers to write a report on the state of German radio/radar and general electronics capabilities and, especially, on any new ideas or technology that they may have developed of which we were not aware. To my great surprise, they offered to give me, for that purpose, the rank of a Wing Commander in the Royal Air Force. Here I was, just 25 years of age, given the dignified uniform with three broad stripes on my sleeve, of a Wing Commander, the equivalent rank of a colonel in the Army. This was to give me the authority to command whatever facilities and access I required as I would travel through occupied Germany. I remember presenting myself to my parents in the uniform of a Wing Commander, and they could hardly believe it. I also remember my awkwardness at having to acknowledge the salutes from all the Air Force personnel as I walked through streets of London!

When I got to Germany, it certainly was an enormous advantage to have the high rank of a Wing Commander. Every British or US military center would provide accommodation, food, and fuel for me and my delegation, traveling in a large military four-wheel drive vehicle with a driver and three scientific appointees to assist me. A rough plan for my trip had been mapped out for me, but I was allowed to make any changes as I saw fit and as I gathered more information.

I went to Berlin first, having to be cleared for part of the drive by the Soviet army, who controlled the western approaches to the city. Of course I knew Berlin well, having lived there for a few years, from 1930 until 1933, and having visited it several times until 1937. The first impression was the enormity of the devastation. Some of the major streets which I knew well that were quite close to where I used to live did not have a single house standing. The street where we had a house for 3 years just did not exist any more. I walked around there among weeds and bushes that had grown up covering the rubble of buildings. I could not even find anything to identify the house in which I had lived.

I went through the various centers. My staff had announced my request for the appearance of certain personnel. We had very formal meetings at which I would sit at the head of the table and, having been told to be entirely formal and correct, I would go through my list of questions prepared for that meeting and my team would note down in detail all of their replies. It was all conducted in English, and if any translation was needed for anyone, it was provided by an appointed translator. I did not give away that I was completely fluent not only in German but in quite a variety of German dialects. South German dialects, which are almost the same as Viennese, I remembered from my days as a small boy in a Vienna school. But here I was the English Wing Commander and my authority would only be diminished if I departed in any way from that appearance. Berlin, divided between the Americans, the French, and the British on one side and the Soviets on the other side, was still an awfully rough place in January 1946. Fights between US and Soviet troops were not rare. The German civilians in Berlin were under a curfew for the hours of darkness, but they did not obey it, even though they risked being shot on sight by the occupying troops.

It was in these circumstances that I had a terrifying experience. Walking home one night with one of my aides from an officer's mess to our quarters, through the empty, dark streets, I was accosted by an American enlisted man. He was tall and lanky, sounded Texan to me (as was verified later), and he asked me for directions to a certain address in Berlin. Knowing Berlin I told him immediately that this was quite far away, much too far to try and walk. While I knew where it was, I couldn't possibly tell him all the turns of the roads he would have to take, so I could really not help him. At that stage, he became very insistent, and he said, "You must take me there." He grabbed me by my tunic and produced a knife that he held in front of my face. My companion, who was a few steps away, did not hear the details of the conversation, nor did he see the knife. He shouted to me, "Go leave him alone, we want to get to our quarters!" Obviously I wanted to leave him alone, but how could I, when I was held with a firm grip by my tunic and a knife was 3 in. away from my face? So I started to walk in the direction I had indicated to him, while my companion started to walk the other way.

It seems the soldier wanted complete compliance with his demands and to coerce my companion to come with us. So he let go of me and chased after him. I shouted immediately, "Careful! He's got a knife!" But it was too late—he had overpowered him, and as I ran to the scene, I saw him bleeding quite profusely and lying on the ground. The soldier meanwhile was ready to attack me and jabbed his knife in my direction as I approached him. There was no one else around, no military police, no vehicles, no pedestrians, nothing. If I ran he might have a pistol and shoot me. If I left my companion, he might bleed to death. At any rate I did not have the choice because the soldier continued to attack me, and I assumed that since he knew he was in deep trouble if he was arrested for an attack on allied officers, he could just attempt to murder us and flee.

I think he was not very skillful at what he attempted to do, for he lunged out at me with his knife in exactly the same way many times in succession. I was to find out later how accurately he had lunged and how accurately I had avoided his attacks, because I had three places where my tunic had been sliced with what must have been a very sharp knife, but without a scratch either on my skin or even on the shirt underneath the tunic. It was the repetitiveness of his attack that gave me my chance, and I abruptly decided to grab his wrist, bring up my knee and wham his arm over it. I could feel and hear his wristbone breaking, and his knife fell to the ground. Instinctively, at that stage I picked up the knife, knocked him on the ground, threatened him with the knife, and, not knowing whether he had a firearm on him, I had to bash him until he was unconscious. An unsavory experience, but the best I could do to go to the rescue of my colleague who was bleeding and possibly to save my own life.

I left the soldier lying in the road and took my colleague to the nearby US military police headquarters. They whisked him away to a hospital to be given a blood transfusion and to be looked after. Then I had to make a deposition (a statement) about the occurrence, and the military police then decided to take me first to the scene of the event. There we saw nothing more than two pools of blood, one in the location of my struggle and the other where my companion had been attacked. From there we drove to the address that I remembered, to which my assailant had wanted to go. We drove there, several miles away, and arrived at a large apartment building in quite a miserable district of Berlin. The two big US military policemen knocked at the front door—it was 3:00 a.m. by then—and a janitor opened the door. I did not know how we would ask for the correct apartment, but it turned out to be unnecessary. The man who opened the door said immediately, when he saw the military police, "He is in apartment number so-and-so, on the third floor." The two MPs just pushed the door open, walked up to that floor, and hammered at the door. The door was opened by an elderly lady who said in German, "But he is not here." The MPs forced the door open and walked in and said, "Where is he?" She said meekly, "In there," and pointed at one of the doors from the hallway. They opened the door and there he was, his arm bandaged, lying in a stately bed, the other half of which was occupied by a girl, stark naked when the covers were rudely ripped away by the military policemen, I presume the daughter of the elderly lady. "Get up, get dressed!" was the command to him, while the shivering German girl tried to conceal herself with a blanket.

The old lady brought in a dripping wet uniform, which had apparently been washed, but still showed signs of the bloodstains, some of them no doubt from my companion's cuts, and others probably from minor wounds I had inflicted on him after I got the knife. He had to put on this dripping wet uniform in January in Berlin, with a broken wrist, tied up with bandages, and come down into the jeep to be taken back to the Military Police Headquarters. It was freezing cold, even in my dry uniform, but what I remember chiefly about my drive back was the fear that the soldier, now incensed beyond belief, would try and go for one of the pistols so prominently displayed on the hips of the military policemen, and shoot me or them with it. I sat there absolutely poised all the time, to prevent any such attempt. Nothing happened. We got back safely; he was handcuffed, even over his broken wrist, and put in some cell. I had to spend the next day writing a detailed description of the entire occurrence. My companion was meanwhile recovering at a military hospital in West Berlin, having received a blood transfusion, and having his hand sewn up. Two days later he was released and joined my expedition.

It is difficult to foretell which abilities will become important in one's life. Certainly here my training in street fights, learned originally in brawls with the Nazi ruffians in the streets of Berlin, had been important and had probably saved my life and that of my companion. The occurrence could easily have become merely one of that large number of "accidents" that occurred in those days in the darkened streets of Berlin every night. I do not know what the result of the trial was to which the soldier was subjected. All that I remember was his last words to me (leaving out the expletives), "You little Churchill, I will get you as soon as I get out." For a year or two I was a little worried, tried to keep my address out of any reference books, keep a low profile (as they say nowadays). But I never heard of this man again.

When my companion was discharged from the hospital, we continued our travels to various scientific and technical centers. An interrogation meeting in Munich with two German technical experts proved rather amusing. I had, as usual, entered the room with my team sitting on one side of the table, the two Germans on the other, and I was to seat myself at the head of the table and commence the interrogations. As I asked my questions from the list in front of me, I heard every now and again a little whisper between the two Germans who were sitting immediately on my left. I gave the appearance of not noticing, but in fact heard the remarks perfectly well, even though they were in the broadest Bavarian dialect. Surely the interpreter, sitting on the other side of the table, could not possibly hear, let alone understand them.

It was normal practice at the beginning of every meeting to read aloud a statement that the persons who were to be interrogated must answer every question to the best of their knowledge, must not conceal any information, and that dire penalties were attached to any infringement of these rules. However, among the whispers was some discussion whether they should or should not reveal the work done in some other establishment. I merely said, in English, "Please conduct your conversation with us in English." They replied that they were sorry that they had only discussed how to phrase their answer correctly. When the meeting was over and as we were walking out of the room, I played my trump card. I said to them in my best broadest Bavarian that I already knew about the work in that other establishment and that there was no use trying to conceal it from us. I shall not easily forget the expressions on their faces! The Germans generally had a high regard for all the British intelligence activities, perhaps excessively high, but here it seemed to be confirmed when they saw a Wing Commander of the Royal Air Force being an expert in electronics as well as in German dialects.

While we were in Munich, I heard some of the horrendous stories of crime and corruption that had grown up, thriving on Black Market sales of US supplies directed to the so-called PX stores, where the US Army could purchase all kinds of things that were not generally available. Cigarettes were the major item and were the general currency in Germany. Liquor, foods, some clothing, all these things which hardly existed in the German shops, were being sold by gangs composed of army personnel and some Germans. Nightclubs and prostitution rings were set up, apparently with enormous commercial gains for the gang leaders. Whole trains had been held up and the PX supplies unloaded into stolen military trucks to be sold on the black market. In the Munich area, I was told that one US Army sergeant had become a multimillionaire, that he controlled many officers and men of the occupation army, and that he could do essentially what he liked. No doubt he had a wonderful career in front of him in the Mafia when he returned to the USA!

After leaving Munich, we had an appointment 2 days later in the vicinity of Salzburg. We drove through the mountains contemplating how to spend our two free days, and to our pleasant surprise, an opportunity offered itself. A small mountain road had a sign in German "Vacation Home for Waffen SS" (the Hitler elite troops). Then underneath was a handwritten notice just saying US Army. I thought this was a good place to put up for the night and maybe stay another day. If indeed US Army personnel were there and there was space available, they would be obliged to provide accommodation. So we drove up the mountain road, and when we arrived on top, in beautiful warm, early spring weather, what we saw was a charming modern building and three very good-looking girls, lying on the sloping roof sunning themselves.

We introduced ourselves and they replied, in broken English, but not with a German accent, they would go and fetch the boss. The boss, a sergeant in US Army uniform, came and invited us in, promised us accommodation and food, and said that he was sure that we would like it there. We saw beautifully furnished rooms, a modern kitchen, a number of bedrooms, I would guess around 20, and many unusual amenities. The shelves in the general living room were stacked with bottles of liquor. Several faucets in the living room were equipped with running beer, instead of water. We were shown how to help ourselves to liquor or to beer, whatever we liked. Two other soldiers appeared, as did a fourth girl, and they all sat down with the girls on their laps, beer in their hands, to celebrate our arrival. To what did we owe these magnificent living conditions? I eventually dared to ask. The reply was that they were a unit dispatched to provide vacation and recovery services to US Army personnel. They had taken over the SS building, which had of course been constructed for similar purpose for the Germans. Why, I asked, is the place quite empty? Are there no US soldiers who need any recuperation? "Well, you see," was the reply, "it isn't sufficiently well known that this place exists." How long had they been there? "Oh, several months." Where had the girls come from? They had come from Czechoslovakia. Patton's Army had briefly advanced to Czechoslovakia and the soldiers had returned from there, having picked up these girls. They had apparently then decided to make themselves a nice vacation home and to live there until somebody found out about them.

The person in charge of this little group was a lieutenant who at that time was in Munich, where he went once a week, to fetch supplies, but he would be back the next morning. Indeed the next morning he arrived, his jeep laden with all kinds of goodies: big hunks of steak, more bottles of liquor, a barrel of beer, and an enormous amount of games equipment, much of it quite unsuitable for that location. He greeted us. The soldiers pulled the goods out of his car, and he immediately took out a golf club from among the game equipment, and he lined up six golf balls on the terrace which overlooked a very steep mountainside. With great expertise, he propelled each of the golf balls on probably the longest flight that any golf ball had ever done, so much for golf on a mountain top.

But there were also skis and ski boots, which were a little bit more appropriate for the location. Indeed we fooled around on skis in the vicinity over much of that day and decided to spend the second night there. It seemed that the lieutenant who was said by his men to be "a very good operator" had the right connections to go down to Munich and pick up whatever he fancied. If no soldiers ever came to use his facility, it was not his fault. He was not responsible for management at headquarters. When we left the following day, we were asked not to discuss this place too much, as it was intended to keep it reserved only for quite special cases. I don't know how much longer these four lived in their blissful condition, but I am sure they did not give it up voluntarily.

When my mission was completed, we wrote a long and thick report about the science and technology we had encountered and the information we had gathered, but it contained disappointingly little that was really new to us. The most interesting thing for me had been a wonderfully well instrumented, very large "swimming pool" in which model ships were tested for their hydrodynamic performance. This was an area where German technology was clearly very good, as we had seen in the performance of many of the ships of their navy. Their radar was uniformly poor, and they had really done no serious radar work in more than 3 years. Hitler was not preparing for a long war, and any technology that would take years to perfect was of no interest to him. Perhaps also the U-boat commander was right that if it didn't make a big bang, the high command would not be interested.

I returned to England, had to give up my uniform and my status as Wing Commander and get back to my radar work. With the war having ended, the urgency had gone out of this work, but I was still very interested in perfecting my moving-target radar. Now we know how easy it is for police radar to clock the speed of an automobile to a precision of one mile per hour, or one part in 10^9 of the speed of light. Still, I was very proud when I could demonstrate that I could distinguish fixed targets from targets moving by as little as 5 mph.

Chapter 4
Return to Cambridge: The Citadel of Learning

Remember: learning's but the yeast,
That makes each kind of bread-dough rise:
For it will make the fool more foolish,
just as it makes the wise more wise.
T.G.

I greatly admired Cambridge and its science departments, for I had by now some understanding of the quality of the research work that had been done there. But at the same time, I was quite aware of the stresses and strains in the system, the jockeying for positions, and the cliques that had grown up there. But the good outweighed the bad, and I definitely wanted to return there to do research work.

Fred Hoyle returned to Cambridge shortly after the end of the war with Japan. He considered he had done his job, and he left the Admiralty before he was officially discharged. He was anxious to get back to academic research. Hermann Bondi had followed a little earlier. These two colleagues had academic jobs immediately awaiting them, largely because of their illustrious past in Cambridge. I did not. I now had to find myself a job.

What had I learnt in my years in the Admiralty? Self-confidence, certainly, because of my rapid advancement into responsible positions and my obligations to deal with many people much senior to myself. On the scientific side, my understanding of physics had been advanced considerably beyond the undergraduate teaching I had had, both through conversations with my colleagues and the study of some texts, but mainly by having to think out many problems for myself. Of course, I had also learnt the latest in electronics, as had many others in wartime radar. Furthermore, I had taught myself and discovered much about ultrahigh-frequency sound.

My idea of how to create a job for myself was based on the possible applications of ultrasound to material testing and to medical applications. I wrote a proposal which spelt out in considerable detail how ultrasound could be used to look into the human body, and I knew that one could not only expect to see the contrast between flesh and bones, as I had already seen, but also quite readily that between different internal organs. I described the ultrasonic "antenna" system that I would propose,

Thomas Gold *edited by* Simon Mitton, *Taking the Back off the Watch*,
Astrophysics and Space Science Library 381, DOI 10.1007/978-3-642-27588-3_4,

a system one would now call a "phased array." Contact to the human body would have to be made through water, and I thought in terms of the patient sitting in a bathtub containing this high-frequency acoustic radar antenna system. I submitted this proposal to the head of the Cavendish Laboratory in Cambridge, Sir Lawrence Bragg, a highly decorated scientist and joint winner with his father of the 1915 Nobel Prize in Physics. In my submission, I asked him whether he would be willing to submit it as a proposal for a research grant to the Medical Research Council. This was the appropriate agency to fund such work, and I thought that I would like to work in the Cavendish Laboratory and also that it would be a place that would most readily provide the necessary facilities. The proposal was so submitted. I made a presentation to the head of the Medical Research Council, Sir Edward Mellanby, and the proposal was approved for a 2-year program, covering my salary, equipment costs, and overhead costs for the Cavendish Laboratory. I was delighted and made my plans to move to Cambridge.

Precisely one day before I had actually planned to move, I received a phone call from Mr. J. A. Ratcliffe, who acted as administrator of the Cavendish Laboratory under Sir Lawrence Bragg. Ratcliffe told me that they had decided to turn down the research grant on the grounds that there was not enough laboratory space in the Cavendish Laboratory for this work. I was devastated! This was obviously the most useful thing I knew how to do. It would be a great addition to medical diagnostic techniques, and I really couldn't believe that in importance, it would not compete favorably with many projects going on at the Cavendish at that time. As I look back on this incident, I think I should have clung to the research grant and found myself another laboratory in Cambridge to accept it. As I realized too late, the zoological laboratory would have been just as good, it was not overcrowded and would undoubtedly have been willing to accept me and the research grant. But unfortunately at the time, I did not know the administrative ways of Cambridge, so I just allowed that research grant to be canceled.

Nowadays, this acoustic echo sounding of the human body is a standard practice and of very great value. Many deformations of internal organs can be mapped without the harmful effect of X-rays and without the need to enhance the X-ray contrast by the injection of X-ray absorbing substances. Unborn babies can be examined with magnificent clarity and unusual features identified. Action and dimensions of the heart and of the major blood vessels coming from the heart can be observed (I have had this done to myself and observed the results). It is a wonderful tool, and the medical profession can no longer imagine being without it. The techniques used and the frequency range employed are all much the same as had been in my proposal. The acoustic contact to the body does not require sitting in the bathtub, although this would be perfectly acceptable, but it is now generally done by the application of a cream as an intermediary between the acoustic source and the skin.

My proposal was turned down by Mr. Ratcliffe in 1946. I do not believe that any such work was commenced anywhere earlier than 1955. My ultrasonic view through my hand, in 1945, preceded any other similar observation, as far as I know, by 10 years or more.

The only useful purpose, to which my ultrasound experience was put, was the construction of the memory system for the EDSAC, the first successful British postwar computer. I met Maurice Wilkes the director of the Mathematical Laboratory in Cambridge and promised him a quick solution for this major, and he thought most difficult, component of his machine. It consisted of taking my well-tried mercury delay line design indefinitely often, whatever string of pulses were to be remembered. A batch of 12 such lines was built right away on my assurance that they would work; this system could store some 24,000 μs pulses (or binary digits) and allow access to any stored signal in 1/500 of a second. Perhaps this was an early application of the technique of changing from a degrading analogue signal (the pulses received in the delay line after each cycle) to a definitely precise signal without any deterioration.

It worked when it was first turned on, and Maurice Wilkes expressed his gratitude in the book he wrote on the construction of the EDSAC, his groundbreaking machine, which ran its first calculations in May 1949.

Some years later, Bragg, when I got to know him, asked me what had become of my interesting proposal to use ultrasound for medical purposes. He knew of that, but he did not know that Ratcliffe, his right-hand man, had turned it down. "What a pity," he said. "If I had known, I would have done something about it." If I learned anything from this episode, it was to be tougher and not give in too readily to authority.

My return to Cambridge had to be postponed, but I was still determined to get a job there. Eventually, I was accepted for work on building a giant transmitter tube, a magnetron scaled up from the radar magnetrons, this one to be used for a particle accelerator. For this, the Cavendish had space and in fact quite a lot of space, substantially more than I needed. I worked on this for some time, but I must say, not with any great enthusiasm. It probably was a very harmful activity to my health because I fashioned the cathodes for these tubes out of thoria, thorium oxide, which can be used as potter's clay. I got myself a potter's wheel, learned the technique, had my big bucket of thorium oxide by my side, and shaped the cathodes with my hands. No one realized at the time how harmful the radioactivity of thorium would be, and no precautions of any kind were taken. I am sure the amount of thoria still in my body could readily be detected even now by counters and would probably be far above the government safety standards (but I do not glow in the dark!).

Theory of Hearing

What I really did in my spare time from building the magnetron was to discuss interesting physical problems in physiology—human and animal—with my friend Richard Pumphrey. He was a senior zoologist in the zoological laboratory nearby, and I went over frequently to see him. His major interest was the mechanism of hearing, the function of the inner ear. He puzzled about all the intricate structures that make up this organ and what their detailed function might be. Eventually, he suggested that I should leave my magnetron work and work with him in this field. We applied and received a research grant, again from the Medical Research Council, and I spent a year working with him in the zoological laboratory.

As I studied the extensive literature and the different proposals as to the detailed function of the inner ear, I became convinced that really only the work of Helmholtz, the great German physicist, represented a basic understanding of the function. Of course, there were many things that we knew that Helmholtz could not have known, but they really all fitted well with his basic interpretation. Let me explain the nature of the problem.

The eardrum receives sound, and its vibrations are communicated by a linkage of two tiny bones to another diaphragm that closes a liquid-filled cavity. It is an elongated cavity called the cochlea, usually in mammals wound up into a spiral shape. It is divided along its whole length into an upper and a lower part by a membrane entitled the basilar membrane, which is made up of tightly packed fibers stretching from one side of the tube to the other. The acoustic input is to the upper half of this tube; the lower half is just closed by a flexible diaphragm. The input of sound thus pumps the fluid in the upper half back and forth, and no doubt the basilar membrane, with its fibers, would be caused to flex up and down.

But before going into more detail about the structure, it is worth thinking about the function that is required of it. The incoming sounds that we can hear go up to a frequency of approximately 15,000 cycles per second. Somehow the message about these sounds has to be sent to the brain by means of nerves, and nerves all function by transmitting individual pulses. It is usual for the pulse frequency in a nerve to be related to the intensity of the stimulus, and the highest such pulse frequency of which our nerves seem to be capable is somewhere in the neighborhood of a thousand pulses per second. But incoming sounds of 15,000 cycles per second can be heard by humans and up to 30,000 cycles per second by other mammals whose design of the hearing apparatus and of the nerves is very similar to ours. It is clear that the nerves cannot function simply like a telephone line, where the electrical signals are just a replica of the incoming sound. Nor could the nerves signal the incoming frequency simply by the pulse repetition frequency in a nerve. Beyond that, one can even argue that the central nervous system is not equipped anywhere to deal with frequencies of 20–30,000 cycles per second. Why isn't it, one might well ask. Why can't our wonderful brains do what a simple little strand of electrical wire can do? The reason is that the only electrical transmission that seems to have evolved in the biological systems is that of ionic transmission. Conducting metals do not appear at all in biology. The chemistry of metals is totally outside the range of the chemistry that is employed in biological systems. Ionic conductivity, that is by the transport of charged atoms or molecules in liquids, like the conductivity of a lead acid battery, is a much slower process. The complex details of the conduction of nerve impulses, in fact, imposed still further constraints. It is clear that a different type of encoding had to be done by the ear, where the single channel of the incoming sound, delivered as it is to the input end of the cochlea, is then translated into a large number of parallel channels, and the great number of individual nerve fibers, such that each fiber does not need to exceed its information handling capacity, and yet the sum total of them, to a sufficiently good precision, includes all the information in the incoming channel.

Although Helmholtz could not have understood the problem in quite these terms because the mechanism of nervous conduction was not known to him, he still understood that a conversion from a single channel to a multiple channel was involved because he knew that the nerve trunk to the ear contained a very large number of fibers. He argued, as any physicist would do readily, that the simplest way to achieve such an encoding was to split the incoming sound into its different frequency components. How then could this be done physically? Of course, it could be done by structures that are resonant, one at least for each frequency that can be discerned and sufficiently many to cover the entire band of frequencies almost smoothly. Then what the nerves have to do is merely to identify which structure, and therefore which set of nerve fibers, is excited to the maximum amplitude, and this will identify the incoming frequency. The intensity can then simply be signaled by the way in which nerve fibers normally do it, namely, the pulse repetition rate. Helmholtz saw that the structure of the cochlea could easily be represented as such a system of closely spaced resonators, and this, as well as more detailed deductions he made, is now all known to be correct. Between the 1850s, when he did his work, and 1948, quite a number of other theories had been proposed, but I did not think much of any of them. Quite trivial points which Helmholtz would have dismissed as obvious appear in the literature as if they represented subsequent major advances. Most of what had been written in these 100 years contributed little or nothing. I would certainly have been perfectly happy if I had never read it, but merely commenced my work with the general explanation of Helmholtz, but, of course, with the additional knowledge of nerve physiology, which was available to me.

Like me, Pumphrey was convinced that Helmholtz resonance theory was basically correct, and we saw how to interpret the fine structure of the cochlea as a system of tuned resonators, namely, the large number of parallel strings that span across and make up the basilar membrane, all equipped with nerve fiber endings that look under the microscope like other motion- or pressure-sensitive nerve endings. We already knew that the nerve fibers in the massive nerve trunk that goes to the brain carry the auditory information and that any single fiber in this trunk has its maximum response for a particular, quite sharply defined frequency. This seemed to confirm that the frequency analysis was done already in the cochlea.

From the microscopic appearance, it was immediately clear, as it had been also to Helmholtz, that the input end is tuned to the highest and that the tuned frequency then descends progressively to the lowest frequencies at the far end. Of course, nobody could follow individual minute nerve fibers and trace out where in the cochlea the nerve fiber had originated whose signal he was measuring with his microelectrode. Had we been able to do that, we could of course have mapped out exactly where in the cochlea each frequency was identified. But that was technically too difficult and still is, but I would guess it won't be much longer. I was able to explain to Pumphrey why the input had to be at the end of the highest frequency tuning, having translated the problem of the conduction of sound waves along the length of the cochlea tube into the analogous problem of an electrical transmission line, for which the complete theory exists. The fact that the appearance of the cochlea, with the short thick fibers at the input end and the long thin ones at the far end, clearly confirmed this, giving me more confidence that the outlook was correct.

Pumphrey and I did a lot of experiments, both of a "subjective" kind, where we merely noted what sounds a person could discriminate, and of an "objective" kind, where we learned to measure the nerve impulses in the auditory nerve of a guinea pig or other small mammal in response to various acoustic stimuli. The subjective experiments were those referred to as masking, where a loud signal is present at one frequency and the hearing threshold of a signal at a different frequency is determined. From such observations, the extreme asymmetry was immediately obvious. A high-frequency loud signal does not do much to interfere with the sensitivity to a low-frequency sound. On the other hand, a low-frequency loud signal screens out the high frequencies very effectively. Of course, we plotted very precise curves, and we could compare those with the objective measurements of the single nerve fiber response to one frequency, which also was quite unsymmetrical as the stimulating frequency is moved away from the peak toward the high- or the low-frequency side. These two very different types of observations fitted very well with each other and made clear that a wave traveling along the length of the cochlea, vibrating each of the fibers, would increase in amplitude gradually as it approached the fiber that was most accurately resonant to the incoming sound, and then beyond that, the wave amplitude would drop off very sharply. Many different types of experiments confirmed this basic behavior. No doubt, the nervous system then sorts out what the frequency is by the sharp gradient of the response following the fiber that saw the maximum amplitude.

Each resonator, which we regarded simply as a mechanical resonator coupled to its neighbors, then has a sufficient number of nerve fibers and nerve endings attached to it to signal in the usual manner, by pulse repetition frequency, the amplitude, or intensity of vibration of that resonator. That, it seemed to me, was the simple description of how the system had to work; it looked as if it did this, and we could be happy with this answer. We might have written our paper, giving all the details and the nice way in which the observations confirmed each other, and we could have closed the book on this chapter. We did in fact write such a paper, but I was more inquisitive than that.

I asked if a system of coupled resonators on this millimeter size scale, immersed in an aqueous liquid, could really have such sharp tuning that was required by all these observations. Would the frictional damping not be quite excessive? I worked out approximately what the damping would be, just due to the viscosity of water, and I found that these small structures could not achieve the sharpness of tuning which we and others had clearly seen in the single nerve fiber observations. The discrepancy was very large. No system of coupling the resonators could give them as sharp a cutoff in the response on the low-frequency side as we had clearly seen in many experiments. A physical explanation had to be found to account for the behavior of the structures in the cochlea and with this for the subjective ability of frequency discrimination.

There was another type of observation which had been made by many other investigators before us, but which we had repeated. It was the observation of a phenomenon whose function was quite unclear. Every sound coming into a live cochlea would produce an electric potential between two chambers which just mimicked the waveform of the incoming sound precisely. It was called the "microphonic potential" because this is, of course, just what a microphone does. This microphonic potential followed accurately even the highest audible frequencies, and thus, it was difficult to see how it could have any relationship with the nervous system, which is limited to a much lower frequency range. Why was this microphonic potential generated? It must have some purpose, but it could not be feeding directly into the nerves. The microphonic potential was remarkably strong, and my estimates were that it contained substantially more energy than was in the sound wave that caused it. It seemed to be an amplified version of the incoming sound.

I pondered over these strange matters for several days and was quite unwilling to let the matter rest. With these thoughts in my mind, I went to the weekly physics colloquium in the Cavendish, and it so happened that on this occasion, the colloquium was extremely dull. A dull lecture is like an experiment in sensory deprivation. You are sitting in your seat, you can't leave the room because that would be too rude, you are carefully shutting out the incoming information because you have decided you don't want to hear it, and your mind is now completely free from external disturbances. It was during this lecture that I suddenly saw how all the facts of the case would fall together. The amplifier would have to work in such a way as to oppose the viscous damping of the resonators. The microphonic potential would have to be applied to elements that acted as electromechanical transducers that took electrical energy and converted it back to vibrational energy, to put back a major fraction of the vibrational energy that had been lost to viscous drag. It had to be a system with positive feedback, just as one had learnt to do in electrical engineering to enhance the sensitivity of some receivers. They were called regenerative receivers. Here then was the reason for the mysterious microphonic potential and for its strength, and this was a way in which the mechanical system in the cochlea could really have the sharpest frequency response at the mechanical level that we had determined. I even remembered that each of the resonators in the cochlea is equipped with two different sets of hair cells, that is, the sensory cells that normally constitute a nerve ending, but that I had seen a remark that on careful examination the two sets do not look alike. Was one set perhaps not the sensory element that one had thought, but the electromechanical transducer that I now needed? It would have to be a cell that lengthened or shortened in response to an applied electric signal. These cells were certainly in the correct place in which such an action would do just what I wanted.

With great excitement, I ran from the Cavendish in Free School Lane to Pumphrey's lab in Downing Street as soon as the lecture was over and explained to him what I had thought out. He understood it right away, and he agreed immediately that it looked like the right solution. "Write up this idea," he suggested, "and we shall publish two papers in succession, the first a joint one in our two names, giving all our investigations, and then a paper in your name, giving the physical explanation for our observations." That is exactly what we did, but, in fact, I decided to write up a version posing the nature of the problem and my suggested solution as a submission for a Prize Fellowship at Trinity College. One fellow of the college, a physiologist by the name of William Rushton, whom I had known and who, like me, enjoyed discussing intellectual puzzles, immediately encouraged me to submit such a paper, and even though time was very short because the deadline for submission that year was only a couple of weeks away, he thought I had a good chance.

Well, I worked feverishly for these 2 weeks and produced a paper of about 50 typed pages that explained in some detail how I thought the mechanism worked. The actual mechanical vibrations in a live cochlea, with a feedback mechanism working, would indeed be as undamped as all the experiments had suggested. The sensory nerves would work in the usual way and just detect the displacement amplitude and then signal this by pulse repetition frequency in the usual way, while of course, it was just the identity of the fiber or fibers that contained the information as to the frequency. It was the "series to parallel" conversion of the incoming information, translated into the code with which the nervous system could work in the normal way.

The paper contained several clear predictions that I then also put into the published paper, predictions of what might be observable or become observable if this theory was correct, even though we had not been able to make the observations. One was the prediction that such a positive feedback system could quite easily flip into self-oscillation and that, in fact, it had to have a careful control to prevent that. Therefore, I surmised that there should be pathological cases where it did go into self-oscillation, which would imply that the ear would produce a sound that one could hear from outside. Most likely, this would happen to just one or a narrow group of resonators, and a sound coming out of the ear would therefore be a well-defined clear frequency. It was in the upper range of frequencies that the feedback action had to be the strongest because the damping would be the most severe, and therefore, most likely a high-frequency sound could be expected in such cases. We had tried, I stated, to induce this phenomenon in ourselves but had not succeeded. A "ringing of the ear" called tinnitus is of course well known as a subjective phenomenon, but that can just be due to a sensitive nerve ending firing spontaneously. That happens commonly in all kinds of sensitive nerve endings in other organs. It was the phenomenon of "objective" tinnitus that we were looking for, where an actual mechanical oscillation was generated that would produce external sound. Although we had not been able to produce this phenomenon at an observable level, I suggested that one should look for it among people who suffered badly from tinnitus and that perhaps in some cases, objective tinnitus was involved. This, it seemed to me, would prove the case completely since an ear not equipped with a positive feedback system but only responding passively to incoming sound waves would have no way of producing such sounds.

The other clear prediction I put in the papers was that one set of the hair cells would be found not as sensory cells, but instead would be cells that responded mechanically to electrical signals (electromechanical transducers). Both these predictions have now been demonstrated in experiments.

There were a few other points that also had clear explanations in these terms, in particular, the various effects that were due to a nonlinear response at high sound amplitudes. It was pretty clear that the electrical feedback system would have no need to be competent at extremely high incoming sound levels and that it would therefore save itself the trouble of perfecting it for that case. The feedback system would thus be too weak to maintain a response proportional to the incoming sound levels at high intensities. Several observations had indicated just such a nonlinear behavior, which would not have been expected of a purely passive system. The feedback system would not only give the high-frequency selectivity of which we were capable, but it would also greatly enhance the sensitivity at low sound levels, and possibly, this was the main reason for its adoption. If one wanted to make a high-sensitivity receiver, one would first have an amplification of the signal and only then the detector—in this case, the nerves. It was a well-known principle in radio engineering, but I don't think it was well known among people who worked in the theory of hearing. As is often the case, bringing information from a different discipline helps to solve a problem.

My fellowship application was successful. I received the Prize Fellowship, and I was now a Fellow of Trinity, which meant that I was financially independent (though by no means wealthy!), and free to choose whatever I liked to do for the next 4 years. Mine had been the shortest successful Fellowship dissertation in any subject other than pure mathematics, where dissertations are normally very short. A Prize Fellowship counted for more than a Ph.D., and so I gave up any thought of working for a Ph.D. Some years later, Cambridge University granted me the higher degree of Doctor of Science (Sc.D.), for which one does not have to do any additional work because it is awarded for peer-reviewed published work of the highest standard.

Pumphrey and I published our papers in *Proceedings of the Royal Society (B)* in 1948. I thought at first that with my solution of the problem with which I had evidently been successful in some Cambridge circles, I would now see the expression of some interest on the part of other workers in the field. I was indeed invited to give a lecture in London to the Medical Society of Neurologists and Otologists. I remember the scene of a large lecture hall filled with elderly gentlemen for the most part, and, appropriately enough for the society, some displaying a nervous tick and some listening with great big ear trumpets in their ears! Electronic hearing aids already existed at the time but evidently did not yet have the approval of the medical profession. As I looked over the audience during my lecture, I could see only blank faces. Polite applause at the end, no questions, and it was pretty clear to me that no one had the slightest idea of what I was talking about. I do not think that they knew the nature of the problem to which I was suggesting the solution.

I then went to visit a number of the individuals who were working specifically on problems of the theory of hearing. Prominent among them was a Hungarian by the name of von Békésy, who worked at Harvard University in the USA. I had an invitation to go over to the USA to work in the National Bureau of Standards in Washington for 2 months, and I took that opportunity to see Békésy. I had long discussions with him, and it became clear that he was very far from accepting the sophisticated theory I was proposing. He had made delicate observations of the vibrations inside the cochlea and found only a flat frequency response, a very heavy damping of the structures. The structures showed some resonance, and a change of frequency shifted the position of peak response, as expected, from a high-frequency response at the input end to a low-frequency response at the far end. However, the response was nowhere near the sharpness that our experiments had indicated. He knew, of course, of the sharpness observed in the single nerve fiber measurements, and in fact, his collaborator at the time of my visit was Dr. Galambos, who had been the first to make such measurements. How did he explain that? Well, he said, there must be some mechanism in the system preceding the location of the single fiber measurement, where the response of a long stretch of the cochlea was registered and the peak point of that response was identified, and then this must be coded so as to give the sharp frequency response in the single nerve fiber experiment.

I had many objections to such a viewpoint. How accurate would the measurement of amplitude have to be so that the maximum response point could be determined with the observed precision? That accuracy was far higher than any subjective information about intensity. Then, what about the recognition of one frequency when it is present in the midst of many others? There would be a combined maximum that would be registered, and we would hear not the individual set of frequencies that we can identify when a chord is struck on the piano, but we would instead judge it as some intermediate tone. von Békésy brushed all this aside and just said that such nervous mechanisms may be enormously clever, and he just wasn't willing to debate this any further.

I asked, had he ever done these observations on the *live* cochlea with the microphonic potential functioning? The answer was no. He had worked on dead cochleas because of the complexity of the instrumentation required. My response was "Of course, you would not have the sharp tuning and you would just see how poor and inadequate a cochlea would be without the positive feedback mechanism."

Békésy also relied heavily on some models he had made, on the scale of tens of centimeters, with rubber strings and rubber sheets, and he claimed that the experiments with these models represented accurately what he had seen in the microscopic observations. I asked immediately whether in scaling up to the large size he had adjusted the various parameters with the correct laws in mind that would apply to such a change of scale. It turned out he had not, and he was really not aware that one could or should apply laws of dynamical scaling to this problem. But just as in experiments with aircraft models in wind tunnels, or with ship models in the experimental ship towing basins, it was of course vital to get the scaling laws right, before anything could be deduced from model experiments.

These and several visits with other people in the field persuaded me that my work, though published in *Proceedings of the Royal Society*, would fall, so to speak, on deaf ears. I was quite convinced that I was right, but I was also convinced that I would get nowhere if I continued working in this field. As a young man, I clearly required myself to have some acknowledged successes, and I was not going to get it here.

It took 40 years, in fact, before it became clear to another generation of workers in the field that my theory had been right. Over the intervening years, a number of the predictions I had made had come to be observed, in most cases, by people who had no knowledge that these predictions existed or what they implied. A family was discovered where father and two children suffered extremely badly from "objective" tinnitus, to the extent that one could plainly hear the sound coming out of their ears. The baby of that family had one high-pitched sound coming out of one ear, a different pitch out of the other. How was this to be explained other than by the positive feedback theory? The doctors trying to treat the problem, since it interfered badly with the hearing of these subjects, thought it might be due to a whistle generated by the blood flow through some malformed blood vessels in the inner ear. They treated the patients by giving them medication that decreased the blood pressure, and they thought their theory confirmed when the tinnitus stopped at a blood pressure level at which they had nearly killed the patient! Many functions had also stopped in these, by-then seriously anoxic patients, and it was not surprising that this particular function was also interrupted. Of course, on physical grounds, it was absurd to think that blood flow through small blood vessels could make loud high-frequency sounds. But it still meant that despite this clear demonstration and the prediction of it that I had made, it was not recognized as having anything to do with my theory.

Several other items came along, the most important being that by a very sophisticated modern technique, it had become possible to observe the actual physical vibration in response to sound in a live cochlea and to determine how sharply tuned this response was, as the applied sound frequency was varied. The curves so measured matched perfectly with the single nerve fiber observations and thus demonstrated, contrary to what Békésy had proposed, that the sharpness of response was really present in the mechanical movement within the cochlea. But again, even this observation was not seen to confirm my theory. It was very frustrating. Perhaps, I should have written another paper, pointing out how these observations confirmed my theory, but then "I told you so," papers do not go down very well. If someone else had done it, it would have been fine.

Next came the observation that one set of the hair cells was seen under electron microscopy to have the structure not of sensitive nerve endings, but instead of "kinocilia," that is, little hairlets that make movements under electrical stimulation. Papers were written expressing surprise and questioning what the purpose would be of having kinocilia in the ear. I was of course working on totally different subjects and did not follow the literature on hearing; but I had an informant who worked in the medical field and who understood my theory and kept sending me the new items as they came along. Most of those working in the field, however, never looked back that far in the literature, and my paper had been totally forgotten. Eventually, however, to my great surprise, I was contacted by a Dr. David Kemp, who worked at a hospital in London on problems of hearing, and he knew of my work, as it was then, of 40 years earlier. He had succeeded with sensitive microphones in observing sounds coming out of the ear, following an input sound: evoked objective tinnitus, as he called it. He had somehow found out about my publications, and he knew the work in the intermediate years. It seemed clear to him that we had the proof of the positive feedback theory of hearing!

I was invited to a conference on the subject, specifically entitled "The Active Cochlea." The background of the story was presented. I was asked to speak about my work of 40 years ago, and I was shown at this conference quite a number of other data that confirmed my story. The most entertaining of them was a movie taken through a microscope, where a number of the "kinocilia" hair cells had been taken out with great delicacy from a mammalian cochlea and were still functional in the liquid on the microscope slide. The experimenter had applied a current through this liquid, and, instead of just modulating it from a signal generator, he had the wit to modulate it with the electrical output of the recording from a rock band. Of course, he presented the sound, together with the picture, and here, we could all see those little cells popping madly in rhythm with the rock band's music!

One of the persons present at the meeting was the hydrodynamicist Sir James Lighthill, who had contributed some detailed work on the action of the cochlea. He later reported on this conference in the *Journal of Vibration and Acoustics* (Vol. 113, January 1991) and included the following passage:

> Last year (1988), many specialists on cochlear mechanisms from all over the world gathered together in England at the University of Keele for prolonged and stimulating evaluation of this idea (later published under the able editorship of Wilson and Kemp, 1989). A special guest of honor at this meeting was that very wide-ranging scientist Thomas Gold who in 1948 won a Fellowship of Trinity College, Cambridge, with a dissertation arguing that the cochlea must incorporate such a mechanical feedback system. In the late 1940s, regrettably enough, hearing specialists had shown themselves unready to take seriously Gold's revolutionary views, and the resulting discouragement led to his moving into other fields such as astrophysics where his creative influence has, of course, been profound. We were happy at Keele to make some retrospective amends to Thomas Gold by displaying an impressive body of evidence to the effect that he had been right all along.

It was nice that someone had looked back that far and found my paper of long ago and connected this with the modern findings. But there had been no continuity; no one had done any research in this field to prove or disprove the predictions I had made and the arguments I had advanced. If I had not had the advantage of knowing personally such people as Pumphrey and Rushton, I do not suppose that I could even have gotten my work published. There was no one else working in the field at that time, or indeed for many years after, who would take this theory seriously, or do work to demonstrate some of its consequences and predictions. Of course, had I been able to submit this to a group largely from outside the field, such as physicists or radio scientists, I would have had little difficulty in persuading them that very probably I had hit on the correct solution. But that is not how the system would work. The peer review system for publication or for presentation at meetings would have called entirely on those working just in this subject. There the probability of a favorable response was evidently virtually zero. The peer review system to which we are so committed is just not set up to take note of anything that is outside of the mainstream of thinking of the day.

The Ratio Club

There was one group, however, who were eager to hear my story and who understood it very well. It was a scientific society put together by John Bates of the Royal London Hospital for the discussion of a subject called "cybernetics." The dictionary defines cybernetics as a term used by Norbert Wiener to refer to the general analysis of control and communication systems, both in living organisms and in machines.

I was elected to this club because Bates, the organizer, had heard of my theory of hearing, which fitted well with the subject matter. I explained in detail there and was very glad that at least in this small but very educated circle, I found much interest and response. One other member of the ratio club was indeed William Rushton, who had been influential in getting me elected to the Prize Fellowship at Trinity (Fig. 4.1).

Harold Shipton. John Bates. W.E. Hick. John Pringle. Donald Sholl. John Westcott. Donald Mackay.

Giles Brindley. Tom McLardy. Ross Ashby. Thomas Gold. Albert Uttley.

Alan Turing. Gurney Sutton. William Rushton. George Dawson. Horace Barlow.

Fig. 4.1 A group photograph of the Ratio Club dating from about 1951. It was a small informal dining club of young psychologists, physiologists, mathematicians and engineers who met to discuss issues in cybernetics

There were several other members in the club who had great influence on the intellectual life of the country. I will mention just two name: one was Horace Barlow, great-grandson of Charles Darwin, a sense-organ physiologist who had become a good friend of mine, having also been elected to a Trinity fellowship, and the other being Alan Turing whom I knew from his visits to me at my war time job in the Admiralty because of my design of the memory devices for computers. I knew that he was concerned with computers, but at that time had no idea how important his work had been for the conduct of the war, nor did the other members of the club. When his wartime work was finally revealed, it became clear that he shared much credit for the winning of the war. With a computer he had designed, he had broken the German code used for communication with their submarines and airplanes, which gave our government advanced knowledge of great importance. The German High Command had considered the very sophisticated code as unbreakable and continued to use it, and Churchill had been very careful to avoid any responses that would give away that knowledge.

During my time in the zoology laboratory with Pumphrey, I was also regarded as the physics adviser to other members of the department, and so one day, Sir James Gray, the head of the department, asked me for advice, and this led to an amusing little episode.

The Eel and the Helix

Sir James was concerned with the many different mechanisms that animals use for their propulsion: not just walking and jumping but also swimming, wriggling, rolling, and all the many things that have been invented in the animal kingdom. On this occasion, he was concerned with the sinusoidal motion that, for example, a snake makes in moving through the grass or an eel in moving through the water. The snake can choose almost arbitrarily what shape it will adopt and how it will alter it in order to glide forward. Suppose the snake puts itself into the shape of a sinusoidal curve, and then proceeds to change the shape to similar sinusoids, but with a different phase relation of the curve relative to the position of its body. Obviously, if it chooses a succession such that its head is in progressively earlier phases of the sinusoid, then it will slide forward. Sir James had, of course, been aware of this. What I added was that the corresponding motion in a liquid would have another requirement in order to work. That requirement was that the motion would have to be one at a high "Reynolds number," which would imply that the inertia of the water was an important matter, more so than its viscosity. A high Reynolds number would require the animal to be fairly large and to move fast. A very small animal trying to swim very slowly in water by wriggling of this kind would be very inefficient, though there may be no other techniques that would be much better. But for animals the size of snakes or eels swimming in water, it would work fine. Professor Gray then discussed that, of course, in water, you could do equally well with a vertical snaky motion. Unlike snakes, he told me, eels have the musculature for doing it either way, and so far as he knew, an eel could be made to swim with a sinusoidal motion in a vertical or a horizontal plane.

All right, we agreed, we can try that out by making up a glass tube of several waves of a sinusoid and sticking an eel into it and urging it in some way to make its way forward in the tube. We could hold the tube once so that the sinusoid was in the horizontal plane and once in the vertical plane and see what the eel would do. But then Sir James came up with another suggestion. He had understood that a sinusoid in a vertical plane and a sinusoid in a horizontal plane could be added in three-dimensional space to make up the shape of a helix. He argued, therefore, that if an eel can do it this way and that way, it also should be able to swim in a helical tube.

There I put my foot down. Yes, the helix is the addition of two sinusoids all right, but it is a degenerate case (meaning that it is a special case where information has been lost). An eel in the shape of a helix will fit into a helical tube any way along its length. For that reason, there is no change of shape that the eel can produce that will give it a forward propulsion. While a piece of a flat sinusoid has a definite phase for its front or hind end in relation to the sine curve, there is nothing similar that one could say about a piece of a helix; if they are the same length, they are all the same. This turned out to be a sticking point for Sir James. Was I denying the profound mathematical rule of superposition? We had a heated debate (all very friendly of course), but Sir James just wouldn't have it. The eel would surely know how to do it!

Sir James suggested we put the matter to a test. He would ask his assistant to fashion a sinusoidal glass tube with several periods of the sinusoid, about three complete cycles, and we would make the eel go through this, held with the axis horizontal and the plane of the sinusoid either horizontal or vertical. Then a second tube would be fashioned in the shape of a helix of the same size, same number of cycles, etc., and we would see what the eel would do in that. My prediction was that it would do nothing. Sir James's prediction was that the eel would go as well through this one as through the flat sinusoid. The following Sunday morning, we were to meet in the fish lab with his assistant and with the completed tubes, and we would see.

Well, we did. The assistant pulled a big eel out of the tank and stuffed it into the flat sinusoidal tube until it was more or less all in, with only a little bit of the tail sticking out. And then Sir James said, "Well all right! Urge it to move along." It wasn't quite clear how we were to urge it, but in any case, the assistant took the suggestion and stuck a little pin in the tail end of the poor eel. The eel immediately understood the message, and it zipped through the three cycles of the sinusoid at a high speed and flung himself out at the other end and fell onto the floor. The experiment was clearly a complete success. The eel, once it started to move, developed immediately a feel for the perfection of the sinusoidal tube for his purpose, and that was the kind of motion it liked!

After a brief rinse in the tank, the eel was pushed back into the same sinusoidal tube, which was now held with its plane horizontal. The eel was pushed in, and again, just a slight indication with a pin at its back end and it shot through the sinusoidal tube and crashed on the floor again. Just as we had both predicted, it really worked.

But now came the third and crucial experiment: the helical tube. Sir James grinned in anticipation of success as the eel was pushed into the tube. No mathematical sophistry was to stop the eel from being just as competent in this medium as in the other. I grinned in anticipation of Sir James's surprise when he found out otherwise. The laboratory assistant had no preconceived ideas about the problem and proceeded as before. When the eel was all stuffed in and only a little bit of the tail was sticking out, Sir James again said, "Well, urge him along." The pin was brought back, but the eel seemed to take no notice of it. "Perhaps he had got used to the pinpricks by now," Sir James said. "Urge him by stronger means!" The assistant's inventiveness was great, and he pulled out his cigarette lighter, and he lit it underneath the tail end that was still sticking out.

This indeed urged the eel to do something. But it was not what Sir James had expected. What the eel did was to make a tight little kink in its shape just behind its head, thereby jamming himself in the previously loosely fitting tube. It then proceeded to shorten itself, which eels apparently can do, and thereby withdrew the tail some distance into the tube. There it stopped.

Now Sir James was looking a bit perplexed, but he said, "Find some other way to urge him along." The assistant took a bit of wire, heated it over his cigarette lighter, and poked it into the back end of the tube at the eel's tail. Now, the poor eel had already contracted himself to his minimum length, and he couldn't do anything in immediate response. But what he did was to make a tight kink near his tail, let go of the kink near his head, and stretch himself to his maximum length. He then made the kink near his head again and again withdrew his tail. As the same incentive for motion was kept up, the eel kept progressing by the same technique, stretching and contracting. It was a very slow and laborious way of moving. Eventually, we got his head a little way out of the far end, and the assistant grabbed him by the neck and pulled him all the way out and put him back into his tank where he swam happily by means of his sinusoidal motions. All that Sir James said was "Well, the eel finally made it through the pipe." I think he realized it was not quite in the way he had expected. I discussed all this later with the hydrodynamicist G. I. Taylor, who then wrote in detail about such motion and the importance of the value of Reynolds number for them.

The Double Boom of Supersonic Aircraft

One day at lunch in Trinity, Ben Browne, the head of the geology department in Cambridge, introduced me to two visitors from the Royal Aircraft Establishment at Farnborough (the principal government research establishment for aeronautics and aerodynamics). They had a problem there, and he wondered whether I could think of a solution.

The problem, they told me, was the following. It had been observed frequently that the boom, or bang, heard on the ground from a supersonic aircraft above, was not one bang, but two, usually in fairly quick succession. Military planes capable of supersonic speeds had only recently started to fly, and while everyone had expected to hear one boom as the shock wave caused by the aircraft swept over the ground, it was far too common to hear two bangs instead. In each case, only one aircraft was in the sky, and they had not been able to find a reason for this phenomenon.

In fact, some supersonic aircraft had been practicing in the Cambridge region, and I had clearly heard this double-bang phenomenon also. There was usually an interval of one or several seconds between the two bangs. The Farnborough gentlemen confirmed that that was also their experience. They said that some people had suggested that possibly one shock wave was due to the front end of the airplane and the other to the back; but then, how could they ever be several seconds apart when the airplane traveled its own length in a much shorter time? It had been suggested that the two shock waves propagated at different speeds through the atmosphere, but they did not like this explanation, and I would not accept it. The bang that you hear had certainly propagated from the close vicinity of the aircraft to the ground at the ordinary sonic speed. A shock wave would be produced only in the close vicinity of the plane, but for the rest of the transmission path, it would be an ordinary sound wave that would result in the boom. So what else was the explanation?

I remembered that when I had heard the phenomenon, I had also heard another strange thing: the engine noise in the interval between the two booms was always noticeably louder than either before or after. When lunch was over, I excused myself and started back to my lab. I had not gone halfway across Trinity Great Court when the explanation struck me with all the facts fitting together perfectly. I ran back and found Ben Browne with his visitors still there, and I told them, "I've got the answer. It is like this." The boom with a person on the ground will be heard by him, but it was generated at the moment at which the *component of the velocity* of the aircraft *toward* this observer was exactly equal to the speed of sound. The observer will of course hear the boom later as given by the distance and sound speed in the air. It would be at that moment that all the disturbance in the air caused by the aircraft piled up into an instant and not spread out in time, and this causes the boom. Now consider the flight trajectory of such a plane. There must always be a moment when the component of the velocity of the aircraft toward the observer first becomes equal to the speed of sound, and then no doubt it will exceed the speed of sound for a little while, but eventually, even if the aircraft maintains a supersonic speed, the component of that speed toward the observer will again fall below the speed of sound; hence, there will be a second moment when it is exactly equal to the sound speed, and this will generate a second boom.

Ben Browne was delighted with the explanation, and the two visitors said that they would have to think about it and take it back to Farnborough before they could really take it seriously and give the verdict.

We can therefore consider three sections for the flight path of the aircraft. Firstly, there will be a section at which it flies at a speed slower than that of sound toward the observer, and over that section, it will of course generate just the usual engine noise heard in the same sense of time as it was produced. Then when it reaches the sound speed toward the observer, all the noise will be piled up at reception into a very brief moment, and he will hear that as the boom. Then even if it continues to fly at supersonic speed in the air, there will be an interval during which it approaches the observer at a speed higher than the sound speed, and during that interval, there will be three components of the aircraft noise that will reach the observer. They will consist firstly of the continuation of the first type of noise delivered in the sense in which it was produced, but the noise that was produced while it was flying at supersonic speeds toward the observer will also arrive but reversed in time. Then it will come to the point where it again changes to a lower speed than that of sound in the direction toward the observer, and again at that point, the noise will all be shortened up into an instant, and another boom will be heard. During the time when it was flying faster than the sound speed toward the observer, it of course causes all the sound received to be inverted in time and to consist of the three components all superimposed on each other. After the second boom, it will again generate the ordinary aircraft noise in the ordinary time sequence at the observer, but this phase will have started already at the moment at which the first boom was produced. So the supersonic stretch of the flight will deliver three noises, two in the normal sense of time and one in the reverse sense. For this reason, the total sound heard in this interval will be louder than that of the first and that of the third interval.

It is a fairly complicated process, and it may take a little work to think it out in detail. Since in the middle section the aircraft was approaching the observer above the speed of sound, it is of course overtaking the sound that was made earlier and the sound that was made later, and so the sound that it makes during that path of the flight will be heard in the reverse order of time, and the first produced boom will be the second one to arrive, and the second one produced will be the first one to arrive. In those days, the actual aircraft speeds were only slightly in excess of the sound speed, and this will tend to make the interval during which the rate of approach is above the sound speed generally very short. This is why we all heard the two booms in very close succession to each other.

Since Farnborough had not puzzled out this problem, I decided to publish the solution in *Nature* where it appeared on November 8, 1952. There was also an article in the popular magazine *Flight* of October 3 on this same subject. That article contains no mention of the essential significance of the velocity of approach toward the observer as distinct from the air speed and hence does not contain the main argument presented here. My *Nature* article was soon widely accepted, and many correspondents confirmed the detailed observational evidence.

Cosmological Theory

I now turn to something completely different. In addition to the work on hearing, I had also spent a lot of time in 1945–1948 with Bondi and Hoyle, discussing cosmology. Hoyle was still full of enthusiasm to understand what the expanding universe could be all about.

What we all knew and understood was that the law of expansion of the universe that Edwin Hubble had discovered was just the one type of law that would allow universal motion to take place without in any way singling out any particular location. What Hubble had discovered was that galaxies moved away from us at a velocity proportional to their distance from us. If in such an expanding system one were to transfer oneself to any other galaxy, exactly the same law would be seen. It seemed clearly an extension of the Copernican viewpoint, which does not place us in any favored position but leaves us in a perfectly average ordinary location. A universe like this, which did not single out any one location in any way, was said to comply with the "cosmological principle."

It was then the general view that this must mean that the universe was denser sometime in the past and would be less dense sometime in the future. Many persons had assumed quite automatically that it meant that the universe started from a concentrated lump of matter a certain definite time ago, a time that can be calculated from the present Hubble's expansion, taking into account how much it would have been slowed down in its expansion by the mutual gravitational attraction of all the masses. A definitive study had been made of such an evolving universe, chiefly by George Gamow, a type of study which is still in full swing today under the name of "big bang cosmology." I had then, and always have had, great difficulties in accepting this outlook. In this, one makes the assumption that the laws of physics as we know them today applied to all the earlier stages, including the very first at enormously higher densities, and the argument for making that assumption is no better than the excuse. "What else could we assume?" Of course, we would not know whether the laws of physics are independent of the structure of the universe in which we find them. The universe is as unique to us, as the laws of physics. If they were *interdependent,* we had not found out how the laws would have changed in an evolving universe. One might think of laws of physics as something of a greater permanence, greater generality, and of our universe as just a particular manifestation of them. Perhaps, our universe is just like one of those exploding fireworks, and there may be lots of other such fireworks, each following the same physical laws.

It is a possibility, but I do not think a very attractive one. My hope was that the development of cosmological theory would allow us, in the end, to understand that a universe of the kind we see was compatible only with the laws of physics we have observed in it. If one could understand in detail that this made a self-consistent package, then one could demonstrate that one had really understood the situation. To give examples in detail, let us suppose that the density of the universe and the constant in Hubble's expansion law could be seen to have some relationship to some of the physical constants like the constant of gravitation, the velocity of light, or the Planck's constant. And, suppose that by understanding this relationship, one could calculate the value of some of these constants from the observation of some of the others. If this gave the right answers, one could claim to have gained some understanding of the nature of the universe. If the universe was not constructed as such a self-contained package, I was just aiming at something that wasn't there. But in that case, I did not understand how we would ever be sure that we have understood cosmology. Just to work out the consequences of assumptions that seemed to me somewhat arbitrary, and with them extrapolate into the past or into the future is just not going to increase the certainty that we have got it right.

Galileo had already been clearly aware of the absence of any influence of a *uniform velocity* on the internal behavior of any system. The local laws of physics do not reveal any uniform velocity with which the system may be moving. Galileo described it as the fly, flying around in the ship in quite the same way, whether the ship was at anchor in the harbor or traveling fast out at sea. I suppose that before Galileo, people had always thought that they could sense motion. All they were really sensing was the bumpity bump of their vehicles as the horses pulled them over the uneven roads. No one seemed to have thought much about the more basic problem, whether velocity without the bumps would be felt.

Of course, Galileo understood completely that the Copernican theory of the Earth traveling around the Sun (in reality, the theory of Aristarchus of Samos in 200 BC) could only be accepted if one knew that absolute velocity was undetectable locally. His "System of the World" (1632) was dependent on what we now know as "Galilean relativity."

Galileo gave great emphasis to this principle of the absence of any effect of uniform motion. Half a century later (1687), Newton phrased his laws of motion, which of course embodied Galilean relativity as a cornerstone. Much later, Einstein had accepted Galilean relativity as so obviously true that he did not even bother to follow the laborious experiments attempting to prove it once again. But he had to contend with a new problem that had arisen. The velocity of light had been measured, and the famous experiment of Michelson and Morley, which intended to see whether the velocity of light in a vacuum depended on the velocity of the emitter or the receiver, had given a null result. Any measurement of the velocity of light, emitter, or receiver being in quite arbitrary states of motion always gave the same result: 2.997×10^{10} cm/s. It seemed very strange that this would be possible. If light had propagated like a bullet, then, of course, it would be measured to have a higher speed when the emitter was approaching the receiver. If Galilean relativity held, then it could not matter which of the two was in motion, for otherwise, an absolute state of motion could be defined. But Einstein was so sure that Maxwell was correct, and that the vacuum speed of light was a universal constant, that he did not concern himself much with these measurements. He did not think that the measurements were in error, but the theory was. If someone had been able to detect absolute velocity, he thought he would have heard about it (his own words).

In 1905, his special theory of relativity provided the answer to this puzzle which had been plaguing physicists for half a century. He showed that, indeed, there was no conflict: the speed of light in a vacuum could be a constant of nature, and all measurements from systems in any state of motion would give the same result for that speed. Only now one had to attend, with careful operational logic, to the measurements of lengths and times that were involved in any one case. This gave different results from those commonly and really quite naïvely assumed: a meter stick would not be seen to have the length of 1 m when measured from a platform moving at a high speed relative to it, and similarly, a clock would not be observed to tick away the seconds in such a case. When the values that would actually be observed in such circumstances are used, then the measured speed of light would come out the same in all cases. Galilean relativity was still true, even with the observations of this new constant: the velocity of light in a vacuum. In any medium other than a vacuum, there is, of course, a preferred frame of motion, that of the medium. Thus, there is no analogous problem for sound: it propagates at a certain speed relative to that medium. Also, light traveling in a transparent medium has its velocity affected by the state of motion of that medium.

We can do nothing in the laboratory to notice the speed of motion of the Earth in its orbit or any other uniform motion it may have. But, of course, if we look out into the sky with a telescope, we could define a local state of motion which would let us see the universe around us with maximum symmetry: the state of motion in which we would have to be to see the Hubble's expansion as most nearly the same in all directions. That would surely be an absolute state of motion, defined by the existing universe around us. It is strange that locally absolute motion is undetectable, and yet, it has a definition in the framework of the universe.

But when we discuss inertia, the situation is even stranger. Acceleration is *locally* detectable by the inertial forces exerted by matter that is accelerated. If someone runs into the back of your car, you may notice that the back of your seat has tumbled and that you are suddenly in the back seat (as I have experienced). You would have no doubt about the inertial force your body exerted, due to the acceleration to which it had been subjected. But acceleration with respect to what? Empty space has no landmarks, no fixed points, and no clocks. Then what does acceleration mean?

In practice, we know that it means acceleration relative to the motion of any system on which no force is acting. That explains how we can determine whether a system is accelerated or not, namely, by measuring its motion relative to any one of the freely moving systems. It does not matter which of them we chose for this determination; whether it was fast moving relative to the surface of the Earth or stationary, we would get the same answer. But this still leaves the question how the inertial force arises and what defines locally the state of motion that is unaccelerated. If empty space does not know what a velocity is, how does it know what a change of velocity is?

The physicist Ernst Mach (1838–1916) considered this problem and proposed that the acceleration that gives rise to inertial forces is an acceleration relative to some average of the masses and their motions in the universe. It was Einstein who first described this statement as "Mach's principle." This principle gave me one reason for thinking always of an *interdependent* system of universe and physical laws. If this fundamental property of matter, that of inertia, depended on the rest of the universe, then obviously, the architecture of the universe was closely associated with the laws of physics. I had the opportunity to discuss Mach's principle with Einstein in Princeton in 1948, and his attitude was the following: the correct cosmological theory must be such that it embodies Mach's principle. General theory of relativity is not in any conflict with it, but he had not found any way in which he could demonstrate that it, in fact, accounted for it. In his own words: "Possibly we have not yet understood completely how the theory and the Universe fit together."

The influence of the masses on the inertia of any object would have to be some average of the masses involved and a function of their distances and velocities to the object. If this inertial force depended too critically on distance, for example, like the law of gravitation which depends inversely as the square of the distance, then this would be a readily observable effect and would have been well known. But, if it depended more slowly on the distance, perhaps just inversely as the distance and not the square, then the great masses with which we cannot experiment but which exist out there would dominate, and any nearby masses we could manipulate would have negligible effects by comparison.

In these considerations, I came up with the following idea: suppose that the inertial forces had indeed just an inverse distance dependence, one could then imagine that the unevenness of the large-scale mass distribution of the universe would result in slightly unequal inertial forces in different directions. If there was a directional dependence, one could estimate that the mass of our own galaxy could make an inertial force different by perhaps one part in 10^9 when the acceleration was taking place toward or away from the center of the galaxy rather than at right angles to that line. Straightaway, I realized it was a rather naïve idea, but sometimes, quite naïve notions turn out to be important. So I set about trying to answer this by an experiment.

I built two electronic clocks with a primary frequency of 10 MHz, each controlled by the longitudinal vibration of a quartz crystal with the axis placed horizontally. I could then compare their frequencies before and after rotating one of them by 90°. The resonant frequency of the quartz crystal would, of course, depend on the inertial forces that it feels in its vibration, and therefore, turning it would cause its frequency to change, if indeed there was a dependence of the inertial force on direction.

To carry out such a measurement to an accuracy of better than one part in 10^9 was not an easy matter, and many technical problems had to be solved, such as temperature control, freedom from external vibrations, and electrical interference. The two quartz crystals were as accurately matched as possible. Each clock was carefully shielded, and only a small amount of radio power was allowed to leak out, just enough to identify it with a sensitive radio receiver in the room. When both clocks were running, I could then hear the beats in the radio receiver, due to a slight difference in the basic frequencies. I then put one of the clocks on a turntable, so I could gently turn it by 90°. Was there any consistent change in the beat frequency when one clock was facing this way or that? After a few hours of operation, it was clear that I would have identified a change of one part in 10^9, if it had been a consistent effect in the two directions. There was no such effect.

I had been quite prepared to get a null result, but I had done what at that time was probably the most accurate experiment of this nature. The best clocks available were only accurate to one part in 10^7, and I do not suppose that any other measurements of inertial forces had an accuracy even close to that. At least I now knew that my naïve interpretation of Mach's principle was not valid. But this did not change my outlook about the principle. There were many other ways in which it could be working. In Einstein's general theory of relativity, it is clear that the masses of the universe define the speeds at which a clock would tick (an effect well observed in the frequency shifts in light from the Sun and indeed dense stars). If clock rates are determined by the distant masses, why not inertia?

A colleague of mine, Dennis Sciama, became greatly interested in Mach's principle and its consequences. Eventually, I advised him to write a dissertation on this subject as application for a Prize Fellowship at Trinity College. He wrote a very interesting paper and was elected to the fellowship. He continued to have a very successful career in several branches of astronomy.

The Steady State Universe

I had objections to the "cosmological principle" according to which different locations in the universe would be statistically similar, *when observed at the same time*. In relativity theory, there is no universal definition of simultaneity: differently moving observers would measure different time intervals for events taking place at two locations.

Would the universe possess a "universal clock" so that simultaneity at different locations could be defined by the reading of that clock? In a big bang theory, where there is a defined beginning of the universe, one might think that universal time-keeping is provided by the elapsed time since the beginning. But even that is not so. In a universe containing gravitating matter, not quite evenly distributed, the elapsed time back to this big explosion could not have a unique value. Different light paths going through this universe would give it different values. Perhaps that by itself was not a serious objection, and one might say, "So what? There is a quantity there, the absolute elapsed time since the beginning, but we have no way of reading it." This seemed to weaken, if not destroy, the cosmological principle.

The other item I did not like in the big bang theory was the necessary assumption that all the laws of physics were independent of the structure of the universe in which they existed. Clearly, this universe would have changed a great deal, and when we looked into the far distance, we would see the physical processes as they would have been in a much denser universe.

There was another problem at the time, which troubled us, but which has by now been resolved because it was merely concerned with an erroneous estimate of the constant defining the expansion—Hubble's constant. With the value then in vogue, and with the interpretation that this specified the age of the big bang universe back to its beginning from a point, it seemed that the geophysical determination of the age of the Earth made it appear older than the universe. Obviously, this would be a conflict in the big bang theory. This conflict exists no longer since the value of the Hubble's constant indicated by more modern observations gives a greater age for the big bang universe. But the story of such conflicts is by no means over, and we will later come to a serious modern one of similar kind.

With these considerations in mind, I hit one day on the idea that there would be one type of a cosmological discussion where these problems would be absent.

That would be so in a universe that, although it was expanding, suffered no change in any of its large-scale averages. Suppose, I said, that galaxies move apart from each other, and new ones form continuously between them. Like a population in biology, there would be an average age that might not change with time, an average density that might stay constant, and so the physical laws might be expected to remain constant also. Of course, as matter moves apart, new matter would have to be created from which new galaxies would be generated. Created from what? Well, I said, let's have it created just in empty space, some elementary particle at a time, with some definite law of the rate of creation per unit volume. What is so bad about that? Why is it any worse than to have all the matter of the universe created in one event, also out of nothing some long time ago? Some kind of creation of matter is required in any case to allow the motion of expansion (except for an oscillating universe, which periodically contracts and then expands again). But creation was also needed to account for the fact that one sees that heavier elements are built up in stars from hydrogen, and one does not see, nor could one readily understand, the reverse process occurring. In this steady state universe that we were discussing, the matter in any one galaxy would keep building up heavier elements, but meanwhile, new galaxies would form, beginning with the light elements. The average distribution of elements on a large scale over many galaxies in the same region would also remain the same. I referred to this new form of the principle as the "perfect cosmological principle."

This universe must be infinite; there must be no limit to its ability to receive more and more galaxies as they expand into this infinite space. The distant galaxies would have to be accelerating so that when they are still farther from us, they will move still faster, to continue to comply with the same expansion law. Does this acceleration require some new force field to cause this outward flow?

Relativity theory provides perfectly self-consistent systems in which such a motion would take place without any new field of force. One can appreciate this by considering Hubble's uniform expansion law together with Mach's principle. Just as uniform expansion does not single out any location, so it also does not single out any particular state of motion as an unaccelerated one (an "inertial frame"). All locations from which the (statistical) distribution of masses and their motions is seen similar to the observation we would make from here would, according to Mach, define the local inertial frame there in just the same way as here. Let us refer to it as "the locally preferred frame." But the Hubble's expansion law is just such that all points following this law will see a similar aspect of the universe around them. This universe would, therefore, define a "locally preferred frame" at every location, and any matter that is not being accelerated relative to this frame will experience no inertial force. If the masses of the universe flow as described, then there are no inertial forces to be overcome to maintain that flow. This would not be quite what Galileo and Newton had supposed; their discussion would be appropriate only for a small enough region so that all points in it see much the same distribution of the flow of distant masses. But objects at large distances from each other in the cosmological frame would have different "preferred frames." There would be no inertial force and no locally measurable acceleration, even though measurement of the recession velocity between two objects that are distant from each other would imply a continuous increase of that recession speed. But then, if Mach's principle held, all the distant masses that defined the local preferred motion would all be expanding and that local frame would then necessarily show such an acceleration. The perfect cosmological principle, the basis of the steady state cosmology, demanded that no absolute time shall be defined in any galaxy and that average quantities like Hubble's constant shall be observed as a constant from any location.

When in 2002 the observations showed such an expansion so that distant galaxies were not only moving but also accelerating away from us, this was regarded as an unexplained effect in the context of the big bang theories. One had expected there to see a gradual slowing down of the expansion due to the self-gravitation of all the masses. Instead, one saw an acceleration of the expansion, without any explanation for it. In the steady state theory, it was a basic requirement, and had it not occurred, this would have ruled out that theory. In the big bang theory, it was just a strange, unexplained effect.

Then, was there anything in any observations that denied the possibility of such a steady state universe? The creation of matter per unit volume and unit time, for example? Of course, one could work out quite readily what this would have to be, in terms of the approximately estimated mean density of matter in the universe and the Hubble's expansion law. It turned out to be a quantity far too small to be observable in any local observation. Was a gradual generation of matter any worse than a sudden one? At least, there could be a defined rate of creation and not an arbitrary instant at which it all appeared. It would be a defined physical law, imbedded in the time and space of the existing universe.

At the time I thought of this ever-evolving, ever-unchanging universe, I confronted my colleagues Bondi and Hoyle with this. At first, both thought that there would be some simple consideration that would invalidate this kind of a picture. After a few days, they had both warmed to the idea. No simple disproof could be seen. Bondi had found out that such a universe would be embedded in a de Sitter metric, one of the simplest that the Einstein theory could provide, and he considered that this fact alone was worth publishing. He also demonstrated a good fit of Hubble's galaxy counts in the different brightness groups, with the strict condition that the steady state would place on such results. These features certainly seemed worth publishing. I myself was keen to publish the basic notion and some of the philosophical considerations that had lead me to the idea. I discussed all this at length with Bondi, who contributed many new thoughts to it. He suggested that we might write a joint paper.

The first indication I had that Hoyle took it seriously was when one day, walking behind him and a colleague, I heard him say, in response to a question, that the only reasonable cosmology was Tommy's, and he proceeded to explain. A few days later when I saw Hoyle, he said not only that he liked the idea but also that he had now put it into a mathematical form. A new field, the C-field as he called it, could be included in the Einstein theory, and this would do exactly what was needed to create the steady state universe I had wanted. He thought that without a mathematical formulation such as he provided, no one would take any notice of the idea. Would I write a joint paper with him?

I thought about this and discussed his formulation with him in some detail. I certainly appreciated the offer, but I turned it down. I much preferred to publish in the first place the basic notion and not make it dependent on a particular mathematical formulation, which might well not be the only one compatible with my kind of universe. I thought Bondi's outlook was much more similar to my own, and his discovery of the geometry of space that uniquely fitted was a very significant step. So I decided on writing a joint paper with him. Hoyle published his C-field paper by himself, and though submitted slightly ahead of ours, it appeared in print only after ours.

The big bang theory was revived by the observation by Penzias and Wilson in 1964 of a three-degree radiation pervading the universe. The big bang adherents immediately seized this and called it the radiation left over after expansion, from that big bang. Hoyle recalls the conversation between the three of us relating to this issue. He and Bondi were of the opinion, still widespread at the time of writing, but false in my opinion, that thermal radiation would carry the frequency spectrum of its creation, unless it was absorbed and reemitted at a lower temperature by an absorbing body. A thermalizing process would be required to turn starlight into the spectrum of the three-degree radiation. I thought that the sum of all starlight would in any case be represented by radiation defined only by the energy density that gave rise to it. Estimates of that energy density existed and showed a value close to 3°. I took the line that there was no case for calling this background radiation anything other than the sum of all starlight and thus that it made no case for a big bang. Hoyle states that the decision was taken, on a two-to-one basis, that starlight could not account for the radiation, and big bang had therefore got the point. He said he later regretted that decision.

The reception of the idea of a steady state universe was varied. There can be no question that it made an impact. Some considered it all as pure madness, others expressed themselves as strongly in favor. The steady state theory certainly became a major item of the cosmological debate for many years, and I would not be surprised if it were to gain attention in the future. The book on cosmology is certainly not closed, even with the present preoccupation with the modern big bang theory. In my opinion, many steps in that theory are too arbitrary, tailored, and adjusted to make the theory fit the observational evidence. The most basic observations for the big bang theory have not been obtained; those would be the observations that the most distant galaxies we can see would also be the youngest. They should therefore show a smaller content of metals in their spectra, as metal content increases and gets distributed to many stars very gradually over the age of a galaxy. But this is not seen. The observations were described by the protagonists of the big bang as having taught us that in the early universe the development of high metal content proceeded much more rapidly than in our older neighborhood. Similarly, the higher space density of matter in the most distant regions was not identified in any way. I have not seen any of the distant observations that show clearly that they exhibit differences from observations nearby. Although all reports of observations now include already the interpretation of the big bang, such as calling objects young if they show a large red shift, there appears to be no observation that rules out the steady state theory.

A field in which the big bang theory claims success is the explanation for the observed concentration of the light elements. Deuterium (the stable heavy isotope of hydrogen) and the elements helium, lithium, beryllium, and boron are present in much higher proportion in the stars than the element-building processes in the stars would have allowed one to predict. It seems possible that lithium, beryllium, and boron were simply the product of very high energy processes that some matter receives. Helium, which is produced copiously in small stars, could be much more abundant than was calculated, using estimates of the density of small stars like our Sun or smaller, and these seemed to show that the quantities so produced would be too small to account for the observations. But we can only see a small sample of such small stars in our neighborhood in our galaxy, and it may well be that this has led us to assume much too small an average abundance of these stars. A much larger abundance would be indicated by all the galaxies that are detected only by the absorption lines they create on long paths from very distant powerful light sources, when the intervening galaxies themselves are too weak to be detected. If faint stars were much more common than we thought, this and the high helium abundance could perhaps be explained.

The theory of the generation of all the other elements was indirectly related to the notion of a steady state universe. Within that theory, one would have to account for all these elements by ongoing processes since one could not appeal to a dense, hot phase of the universe and since one could not imagine starting with complex nuclei. This directed attention to all that might be going on in the various stages of the evolution of stars as we know them.

Gamow had tried to account for that complexity of all the heavier elements by nuclear processes that happened in the first brief moment of his universe. The matter was very hot and dense, but nevertheless, he could find no way of reaching carbon-12. He also wanted to start, with protons and electrons making up hydrogen; any mix of more complex nuclei created initially seemed to him also too contrived. The addition of protons, one by one, or helium nuclei could not bridge over the nuclear mass numbers of five and eight, for which no stable nuclei seemed to exist. Triple collisions such as assembling three helium nuclei to create carbon-12, for example, could just be ruled out. Whatever circumstances one assumed for the brief high temperature phase of the big bang, the chances of having three helium nuclei meet together within the short time interval defined by nuclear processes, would be much too small to create the abundance of carbon and all heavier elements beyond.

Hoyle together with Willy Fowler and Geoff and Margaret Burbidge looked at all this from another viewpoint: can we see how all the elements could be constructed in a steady state universe, in processes that may be slow, but that are happening all the time? Dense, hot interiors of massive stars could facilitate nuclear reactions, but here, long time spans were available for rare events to become significant contributors to the final mix. Could three helium nuclei be assembled to make carbon-12? Edwin Salpeter had concluded that if beryllium had an isotope of mass 8, even if only for a brief moment, then in the long time spans of a star, this isotope could suffer a collision with another helium nucleus and thus produce the stable nucleus of carbon-12. The nuclear physics laboratory of Caltech under the direction of Willy Fowler established that there would be indeed a short life span of beryllium-8. The triple collision problem, which Gamow had already recognized as a barrier to the building up of the elements from the simplest, namely, hydrogen, was circumvented. The next step in the buildup of the elements would be the addition of another alpha particle to carbon-12, turning it into oxygen-16. The approximate abundance ratio of carbon and oxygen was known, and Hoyle noted that this ratio could be explained only by supposing that a resonance existed in the carbon-12 nucleus, within a narrow range of energy. In summer 1953, Hoyle urged Willy Fowler at Caltech to look for this, and indeed, it was there. A detailed prediction derived from cosmological considerations was confirmed in the laboratory.

Once a buildup of the elements in the stars could get to carbon-12, further buildup by the addition of helium or hydrogen nuclei could make the sequence of the elements up to the vicinity of iron. Beyond that, the heavier elements could be constructed from neutron irradiation, with neutrons produced during the violent collapse of massive stars. The famous papers by the Burbidges, Fowler, and Hoyle explained many of these processes in fine detail, and the fit to the observational quantities, like the relative abundances of different elements and their isotopes, leaves little doubt that these explanations were correct. I regarded this work as one of the greatest importance; for the first time, we knew how most of the many different elements had come into existence.

A serious trouble of the big bang theory is an age conflict that has arisen again. Just as the conflict with the age of the Earth I mentioned earlier, it is now (I am writing in 1994) with the ages of stars in globular star clusters of our galaxy. Some of these seem to be about twice as old as the modern measurements of the Hubble's constant would give for the age of a big bang universe. This time, both measurements seem remarkably secure: the ages of stars in a cluster can be identified very closely by their spectrum and brightness. The fact that one has been able to match the theory of stellar structure very closely to the complex patterns of brightness and color in clusters must surely mean that the theory cannot have any serious error. The distance to globular clusters is known well from a comparison of the spectral Doppler shifts and the transverse velocity distributions that are observed.

The modern observation of the Hubble's constant also seems more secure than any previous determination. Oscillating stars, the so-called Cepheid variables, have long been regarded as the best "standard candles," light sources whose *intrinsic* brightness was known when the period of oscillation could be identified; the *apparent* brightness is then an indicator of their distance. This observation can now be done both with the Keck telescope and with the Hubble Space Telescope, out to much greater distances than before, giving a much more reliable value of Hubble's constant. These observations and their implications could turn out to be like pulling a bottom card out of a tall card house. I am sure that much effort will now go into trying to repair the damage since many investigators will consider it virtually impossible that the large structure that they have built up on the big bang basis could be subject to complete collapse. I do not claim to know what the right answer will turn out to be, but the wider range of possibilities which the steady state theory initially introduced will surely now receive fresh consideration.

How do I now feel about the subject of cosmology, in 1994, 46 years after my contribution to the field? Several of my critics, over the years, have said something like this about my record in science: "It is true that he has quite a good track record, but then he has also made gigantic mistakes—like the Steady State Theory."

Whether that theory will turn out to be totally wrong, almost right, or completely right, I will not be in the slightest embarrassed to have proposed it. In a subject of that nature, any theory that can contribute to the debate and hold the attention of many investigators of the field is important. If it was a gigantic mistake, it was one about a gigantic subject. If it turns out to be right, or to direct future considerations in the right direction, it will have been a success. On the stock market, a gigantic success is measured by the gigantic financial gain that resulted and a gigantic mistake by the loss. But in science, there is no such simple judgment. Most steps are mistakes; just as in biology, most mutations are mistakes. But out of it, all will come a further, better understanding of the subtleties of nature.

Debate with the Cambridge Radio Astronomers

At a conference in the spring of 1951 in London, the radio astronomer Martin Ryle (later Sir Martin Ryle, Nobel Laureate and Astronomer Royal of Great Britain) had presented his observations of some 50 radio sources rather evenly distributed over the sky. Ryle claimed at that time that these must be strongly radio-emitting stars within our galaxy and indeed at rather close distances to us. If they were stars in our galaxy that were rather far on average, then of course they would show the concentration toward the band which is the Milky Way, in which we see the great multitude of more distant stars. This meant, he thought, that radio stars had to be very common so that we would pick up just the nearest ones and only see a blurred distribution of the sum of all the distant ones, just as the sky looks to us with the naked eye: we see individual bright stars distributed more or less equally all the way around us, and we see the Milky Way as just a narrow band of diffuse light.

I was very unhappy with Ryle's discussion of the problem as I did not believe that ordinary stars, even if fairly nearby, had any way of emitting such strong radio signals. In my discussion at the conference, I mentioned that the only type of star that could be a strong radio emitter, in my view, would be a collapsed star of high density, which had concentrated its magnetic fields by its collapse and which would then be a dense object with very strong magnetic fields extending out into its surrounding space. That would be the ideal circumstance for creating a strong electromagnetic interaction and for making intense radio radiation. However, I thought that if these were Ryle's radio stars, they should show time variations in the short times that light takes to cross such sources. Since no such time variations were seen, I did not really believe that these sources represented stellar objects. Instead, I proposed that they represented distant galaxies in which large-scale and very intense electric phenomena were taking place.

If I can digress, a few years later (1963), Cyril Hazard made the first accurate position measurement of 3C 273, the first of a class of galaxies that received the name of "quasars" (quasi-stellar objects), very distant galaxies containing an intense source of light, so concentrated that it appears on the photographic plate like a star. They were at the same time very strong radio sources, and they represent many of the objects Ryle had observed in his "Third Cambridge Catalogue of Radio Sources." Then, in 1967, the same Cambridge radio astronomy group discovered the objects named "pulsars," which indeed were radio stars but much too faint to represent the objects Ryle had noted. These did, in fact, show rapid time variations, and I recognized them immediately as the objects I had discussed as possible radio stars at this conference. I therefore identified them with collapsed, dense stars, "neutron stars" with strong magnetic fields. But more about this later.

Concerning Ryle's "radio stars" of 1951, if the distribution of radio sources seen showed no concentration toward the galactic plane, then one could argue either that they were for the most part so close that the more distant parts of the galaxy made no contribution or that they were extragalactic objects, like the distant galaxies. It was only the in-between case that was excluded by the observational evidence. Fred Hoyle, who was also at the conference, was of the same opinion as me. In the intense debates that followed, it was clear we were out on a limb so far as the majority of the participants were concerned. Ryle said, "The theoreticians have misunderstood the observational evidence." He meant, of course, Hoyle and myself. It was quite a heated debate, and Ryle, a man of intense passion, came away from it, and I felt wounded because his carefully prepared case had received a challenge.

There were several other points that came up in that same debate that turned out to be important. Ryle's explanation of the diffuse band of unresolved radio radiation from the plane of the Milky Way seemed unnecessary to me. My response was that I needed no explanation for this radiation, for it could readily be interpreted as a very small by-product of the cosmic rays, which we thought were largely confined to the galaxy (now generally considered to be the explanation). Distant galaxies might well have a small proportion of very active ones in which processes were going on that were largely absent in our own galaxy. Some distant galaxies had been seen with greatly broadened emission lines, indicating that violent motions were taking place in them which had no counterpart in our Milky Way. There clearly were great differences in the processes that go on in different galaxies, and we thought that this would better account for his radio sources than some new population of stars of a kind that had never been observed in any other way, and for which I could not see any reasonable physical model.

Just a year later, the International Astronomical Union met in Rome. It was there that I saw the extreme emotional involvement of Martin Ryle in his work. I was standing in the foyer of the conference hall, and Walter Baade, a friend and senior astronomer at the great Palomar observatory, called me over to look at some pictures. "Here Tommy," he said, "You'll be pleased to see what we brought from Palomar." What he had was a 200-in. photograph with very good resolution in the position of one of the strongest radio sources known, the source Cygnus A. The plate showed what looked like two strangely superposed galaxies. But Baade had more to show me: a spectrum taken in May 1952 by his Palomar colleague Rudolph Minkowski. The unusual spectrum showed a recession velocity of 9,500 miles per second, meaning the so-called radio "star" was hundreds of millions of light years away. Baade interpreted the photograph and the spectrum as a collision between two galaxies, but added, "this interpretation may be incorrect." However, it was clear that he had a very strange-looking galaxy at the position of the radio source. Baade said that he had read the discussion that had taken place the previous year at the London conference, and he knew that I favored the notion that galaxies were responsible for most of the radio sources. Here was one of the strongest, and it clearly was an extragalactic object.

While we were looking at the pictures, Ryle came in, and Baade called him over in his usual jovial way, saying, "Here Martin, come and see what we brought from Palomar." Ryle came over, had a brief look at the pictures, and heard Baade's interpretation as a pair of colliding galaxies. It was an unfortunate accident that he was confronted with this information just in my presence since he undoubtedly remembered the debate of the previous year, and even though I did not say a word, he knew that I had to repress an "I told you so." Ryle was so upset at the information that he threw himself on a couch, buried his head in his hands, and sobbed. Baade looked at me with a puzzled expression, shrugged his shoulders, and said, "Did I do anything wrong?"

The next round of the debate took another form. Ryle was now convinced that the objects were extragalactic. He was in the "cosmology game" and would now use radio sources as a probe into the most distant regions of the universe to contribute to the understanding of the cosmological picture. Since most of the sources did not have an optical identification, and since one could not determine any spectral shift from wideband radio signals, all that could be done was to plot the numbers of sources above a particular value of intensity against the intensity. By plotting the logarithm of the numbers against the logarithm of the intensity, a simple calculation shows that for a uniform distribution in space, this plot should be a straight line with a gradient of -1.5. This interesting result, as Ryle and his theorist colleague Peter Scheuer pointed out, is true not just for sources that all have the same intrinsic power, but even for sources with any spread of intrinsic power, so long as there is no consistent change of intrinsic power with distance from us. In the steady state theory, there should be no such consistent change, and the plot could therefore make the distinction between this and other cosmological theories.

Ryle's survey a year later showed some 200 sources, and the gradient of this graph was very different from -1.5, more like -1.8. Ryle took this to be a clear disproof of the steady state cosmology, and his claim received much publicity. He presented this with much aplomb at the Royal Society in London. What an entertaining game this was! My view, against which he had fought so strenuously before, was now to be used against our cosmological theory! After these announcements, I visited his group in the Cavendish Laboratory with a request to get more detailed information about the observations. He assembled several of his colleagues around him, ready to answer any questions. One of my questions concerned the precision with which they were measuring the intensity of each individual source. They gave me some approximate figures, but then Ryle said, "Oh, but it really does not matter how precise such measurements are, since Scheuer has shown that such errors would not affect the slope of the graph." He had evidently momentarily confused the demonstration that the sources did not have to be of standard strength to give the result, with the very different conclusion that one did not have to measure them with precision. My response to this was, "Well if the precision of the measurement is not important, why do you have a radio telescope?" The laughter around the room was not received kindly by Ryle, and this, as well as several other little episodes, produced somewhat strained relations. Hoyle attributes the troubles he had over the years in Cambridge with Ryle and his associates, to the fact that Ryle's pride was wounded because we had been right about the extragalactic nature of the majority of the radio sources.

Two years later, Ryle had the "Second Cambridge Catalogue of Radio Sources" with about 1,700 radio sources. When this was plotted, it showed a slope still not 1.5, but much closer to that value than the earlier one. It was clear that the first survey had been substantially wrong, but this time, a much greater precision was being claimed for the new one. Confusion, the inability to separate out the different sources and lumping several together as one, had been a major source of error. I think even now, some 36 years later, this subject has not been cleared up completely. Whether the steady state theory of cosmology is right or wrong will be cleared up, but it does not seem that it will be done by the methods advocated at that time.

Much as I was in opposition to Ryle's claims at the time, I later became convinced that he had made considerable contributions to radio astronomy and to radio techniques in general. I had learned and understood the notions of "aperture synthesis" from W. N. Christiansen in Australia, a highly original radio astronomer who, with a big grin on his face, would explain the most elaborate trickery of radio methods, but who would also demonstrate their application with detailed designs of instruments. Ryle had understood these matters, whether independently or from Christiansen we do not know, but he applied them on a grand scale. From a set of antennas movable over a certain area, his group could obtain almost all the information that a single antenna encompassing the entire area would have obtained. The quality of this approximation depends on the patterns and numbers of positions to which the antennas had been moved in a set of observations of the same celestial object. Radio astronomers could not contemplate building single antennas of the enormous size of the area over which these piecemeal observations could be carried out, and therefore, a great advantage had been gained. Of course, the signal strength obtained in this way would be far less than that which such a large antenna would have gathered, but the angular resolution would be there. It was the successful application of this method to radio astronomical problems for which in 1974 he received a share of the Nobel Prize in Physics.

Electrical Phenomena in the Solar System

One day in 1951 at Cambridge, I was asked to meet with a visitor from Sweden, a Professor Hannes Alfvén, and I was told that I would be most interested in the new ideas he was bringing into physics. Indeed, I was. I talked with Alfvén for many hours, and I came away with a new understanding of the application of the laws of electromagnetism to moving and electrically conducting fluids. He explained that on a small scale, these results would be unimportant, but on a large-scale, conducting fluids, like the ionized gases in the solar system or in the spaces of the galaxy, would behave according to the rules he had outlined. Magnetic fields in such a fluid would move with the motion of that fluid. The magnetic lines of force, previously merely a definition of the direction of the field at each point in space, now assumed an identity. In this approximation, which is good only for a very large-scale and a high conductivity of the fluid, the set of fluid particles that were on one line at one time will continue to be on one line at all later stages of the fluid motion. The magnetic fields would be just convected around by the motion. It was a very simple concept, and it made it easy to see what the consequences of a motion would be for the magnetic fields that were embedded in that fluid. This concept became the cornerstone of a new subject, called magnetohydrodynamics (what a dreadful name for an interesting subject!) or MHD for short.

Many people were not ready to accept these simple rules. I recall trying to persuade Sydney Chapman of them, but without success. He was the commanding figure in such subjects as conductivity of ionized gases and electromagnetism in the solar system. Many of his considerations would have been greatly simplified with the aid of these concepts, and some would indeed have been found at fault. But Chapman insisted that he would work things out "the proper way," even though the cumbersome mathematics often did not allow one to recognize the major features of a problem. Also, he saw that Alfvén was inexact and sometimes in error in his calculations, and that put him off greatly.

Werner Heisenberg, on the other hand, saw all this in a flash when he was in Cambridge and visited me at my house. The only other experience I have of anyone following my explanations (in another field) so effortlessly and quickly was with Richard Feynman. (I was asked once, "give your grading of the intellect of scientists you have met, on a scale of one to ten." My reply, "Feynman, 12.")

Strangely enough, Alfvén himself did not always follow all the obvious consequences of his rules. I visited him at the Royal Institute of Technology in Stockholm, and he proudly showed me a huge vacuum apparatus in which he could shoot an electron beam at a magnetized sphere, with the expectation that this would mimic an emission from the Sun and the creation of a magnetic disturbance on the Earth. I objected and said, "but you are leaving out all the magnetized gases and their motion in the intervening space, aren't you?"

There was so much to be done with these simple notions. One application was to the understanding of the terrestrial magnetic storms and their relationship with the observed disturbances on the Sun. Also, I thought, the great enigma of the origin of the Earth's magnetic field might perhaps be solved. The hardest thing I did in preparing myself for these tasks was to read the two very thick volumes on "Geomagnetism" by Chapman and Bartels. The origin of the geomagnetic field was discussed only on half a page, and they did not suggest any mechanism. However, the effects of solar events on the geomagnetic field were presented in very fine detail.

One thing puzzled me greatly. Most magnetic storms (severe disturbances of the geomagnetic field) had a "sudden commencement," that means, they started not gradually but with a sudden 1- or 2-min increase of the strength of the Earth's field. This was then followed by the main phase of the storm, a wildly fluctuating but generally weakened magnetic field. Many or most of these storms followed a visible solar disturbance with a delay of 2 days or so. How could something that took 2 days to come here have as sharp a front as to produce the 2-min commencement? Why would such a front maintain itself during the 2-day flight time and not be diffused by the thermal motions in the gas? That front had to be a slab as thin as one part in 1,400 of the distance to the Sun or 100,000 km; that is only one quarter of the distance to the Moon. How was this possible?

Shock waves in gases were known as a phenomenon of the interaction of gases at supersonic speeds. But such shock waves had a thickness that depended on the collisional interactions of the gas particles, a few mean free paths in the gas, and in this case, the mean free path was about as long as or longer than the entire distance to the Sun. So, that could not be the answer. What I concluded was that there must exist shock waves that are mediated by the magnetic interaction between the gases, not by particle collisions. At a conference in Cambridge in July 1953, entitled "Gas Dynamics of Cosmic Clouds," I presented this novel explanation for the sudden commencements and the necessity to invoke "magnetic shock waves" in interplanetary space. In response to the intense opposition that this provoked, I replied:

In considering the interaction between the stream and the residual gas one must not restrict oneself to the collision cross section of neutral particles, but one has to consider the much stronger electromagnetic interactions that may occur between the two ionized gases.

It took quite a while before "collisionless shocks" or "magnetohydrodynamical shock waves" became accepted, even though Arthur Kantrowitz, the gas dynamics expert, went to great lengths to spread the word in gas dynamics circles and even set up an experiment in his laboratory to demonstrate the existence of these waves. I had thought that this picture would immediately be clear to the scientists concerned and that the sudden commencement argument was beyond debate. But I was gradually learning that it is very difficult to change an established opinion. Now, of course, MHD shocks are seen to be a major feature in astronomical gas dynamics and in plasma physics experimentation.

Later, I wrote quite a number of papers between 1953 and 1963 on the subject of MHD effects in the solar system. These include the discussion of magnetic storms and their relation to solar events; the cause of solar explosions, called solar flares; the method of entrapment of hot ionized gas (plasma) in the outer reaches of the magnetic field of the Earth, a region for which I invented the word "magnetosphere"; and the effects of magnetized clouds emitted by the Sun on the galactic flux of high energy particles, on the cosmic rays, and on the solar high energy particles, given the misnomer "the solar cosmic rays."

Election of Chancellor of the University of Cambridge

Toward the end of the year 1950, the election of a new chancellor of the University of Cambridge had become necessary, due to the death of the previous chancellor, Field-Marshall Smuts. The Second World War had been over for 5 years; the country had suffered gravely and had not yet recovered too well. Wars, though usually won by Britain, were nevertheless regarded as a thing of the past. By 1950, the Empire was no more. Military glory was of no interest to the young generation. For instance, I recall that films of British military glory, still plentiful then, were jeered and ridiculed by Cambridge undergraduate audiences. On the other hand, the thinking liberal fringe had occasion to take pride in the manner of the dissolution of the power structure of the empire. Instead of waiting for this to be dismembered by a succession of small wars, the British government took the courageous steps to guide the dissolution itself. Churchill's words had been, "I am not here to preside over the dissolution of the British Empire." But instead of Churchill, it fell to Clement Attlee, the prime minister in the Labour government that had meanwhile come to power, to oversee much of this dissolution. The greatest component of that past British Empire had been India. India had won its independence without a major conflict with Britain. In 1950, to many a young man, this was where glory lay, to lead the world from its absurd ways of disastrous power politics to an outlook that tried to predict the inevitable future and then chart the best route to it.

The chancellorship of the University of Cambridge is not an important matter in itself. No chancellor has much direct influence on the evolution of the university. The post is, however, one that symbolizes what is held in high esteem by the university at the time. After all, there is nothing that is more important than what is in men's minds. The fight for chancellor is the fight to give expression to this, to symbolize an outlook, and if this is done by the Senate of a great university, its staff, and its past graduates, it is an important matter indeed.

It is the vice chancellor who is traditionally the person who presides over the organization of the university and with that over the Senate.

Some days before the impending election, the vice chancellor, Mr. Willinck the Master of Magdalene College, called an informal meeting in the Senate House, for a discussion of this election. After that, I heard from various people at the high table in Trinity that "it was going to be Tedder" and then that the majority of the old guard had already fully committed themselves to this decision. Lord Tedder had retired as Marshal of the Royal Air Force. I was by then an M.A. of Cambridge University and thus had voting rights in the Senate. As it happened, I could not attend this informal meeting, but I had in any case no intention of speaking up, thinking that many possible candidates would be discussed by people more senior than myself, that they would be sure to include Jawaharlal (or Pandit) Nehru, the prime minister of India, which Britain had made independent in 1947. I had not reckoned that the "old guard" would take it upon themselves to confront this informal meeting with a *fait accompli*, rather than genuinely sound out the ground to be able to judge the feelings of the graduates of the university. Instead, as I later heard very clearly, the informal meeting was just devoted by the old guard to make the strength of their position so clear that only one candidate would come under serious discussion. The only name besides Tedder that was even mentioned at that meeting was that of Lord Mountbatten (the uncle of Prince Philip and the last governor-general of India).

Tedder had been an admirable leader in the defense of Britain against the German bombing campaign, as the air chief marshal responsible for the fighter command of the Royal Air Force. I had great respect for him, but did not like the symbolic implication of electing him. Pandit Nehru seemed to me incontestably the greatest figure to be honored by the university, being also a Cambridge graduate, and having fought for many years, at great personal hardships, for the freedom of India and then achieved it without a conflict with the British crown.

In the following days, there was much discussion, in which I took part and expressed the view that now was no longer the time for a great university to express admiration for military skills. Since Lord Tedder's fame was entirely based on his wartime career, it would be inevitable that this would be taken as the basis of his election. I would not have opposed the nomination of Lord Mountbatten, for although he had many military honors to his name, his fame was also based on his great achievements as viceroy of India. His connection with royalty would be regarded by some as a contributing favorable factor. Nehru's opposition to militarism was well known, and at the same time, we could be confident that his name would bring out many of the most prominent persons on the Cambridge scene who favored his candidacy and who would cast their votes in the Senate. It seemed to me that he should in any case have been the first choice. A further valuable and genuine argument in favor of the choice of Nehru was the effect it would have in Asia. There, at that time, much anti-European feeling was based on the white man's conceit under colonialism, and the election of an Asian to the most honorable office of a major university, especially in the country that had been the world's leading colonial power, would surely have a beneficial effect. It would show that the change was genuine and that there would be no return to the past.

The steam roller tactics of the old guard had drawn much adverse comment from the younger generation, but it was not confined to that. I soon found that there were many persons of great global reputation, who had heard of our interest in opposing the steam roller. The biochemistry department, traditionally the most forward looking, had a group who desired to sign up on a petition to propose Nehru. When I went to that department, I found out that there was no one who had collected signatures, but many who desired to have their signatures included in any petition. It was at this stage that I decided to take action.

I had discussed the matter before with some of my closest colleagues. Jack Gallagher, a cofellow of Trinity (and many years later, its vice master), had advised me to lie low; there were too many senior people on the Tedder side so that we would just be run over. But then, when we had a solid body of supporters, he must have changed his mind because he became a strong supporter of the campaign for Nehru. Nick Kemmer, also a close friend and a physicist of great distinction, clearly desired to help the same cause. Between the three of us, we collected signatures. This was soon so successful that the process could not be stopped.

Kemmer, Gallagher, and I had put our heads together to produce a leaflet to send out to some thousands of Cambridge M.A.s, mostly out of town. It was a one-page leaflet (known locally as a fly sheet) that contained just the first list of important signatories of the Nehru petition we had obtained. Jack made his rooms in Trinity available for the campaign, and a large number of undergraduate volunteers soon filled the facilities to help with all that correspondence. I am giving here the text of the leaflet.

Election of Chancellor of the University of Cambridge

The Prime Minister of India is, among Cambridge men available for the office of Chancellor, incontestably the most eminent. It is therefore our belief that he should be elected to fill the University's most honorable office.

It is true that several arguments which might have been brought forward against the recent election of the late Field-Marshal Smuts to the same office, may now be urged against the election of Pandit Nehru; he too, is a statesman playing an active part in the political life of a great land within the Commonwealth; he too, lives far away from the University; he, too, has been in his time an opponent of the Crown.

These considerations were not held to weigh in the scale against Field-Marshal Smut's illustrious services to the Commonwealth. It was generally felt at that time that the University's act in offering to a Commonwealth statesman, not born a subject of the King, the highest honor within the University's power to confer, might everywhere be construed as a sign of liberal and imaginative ways of thought and feeling, which it is an important duty of the University to foster. In a world torn by nationalism and prejudice the University then made a clear statement of its belief in the ultimate peaceful reconciliation of differences between men.

If these considerations were valid and counted for anything in the election of the late Chancellor, they are even more urgent in the choice of our next Chancellor. The need for toleration and liberality of thought and feeling has become not less but greater than it was then.

We do not wish to deny or to detract from the deep obligation this country owes to Lord Tedder for his outstanding services during the late war; but Pandit Nehru, as Prime Minister of India, has it in his power to offer to a world distracted by hatred and prejudice services incomparably more valuable and more pacific than lie within the grasp of any other Cambridge man at this time.

We ask members of the University to offer to Pandit Nehru, who is a scholar as well as a statesman, the office of Chancellor as a mark of admiration of his qualities of character and of intellect, and as a sign of our hope and trust in the peaceful reconciliation of the different races and creeds of mankind.

We hope that those who share these views will come to Cambridge to record their vote.

E. M. Butler
(Newnham, Schröder Professor of German)
H. H. Farmer
(Peterhouse, Norris-Hulse Professor of Divinity)
S. R. K. Glanville
(King's Herbert Thompson Professor of Egyptology, F.B.A)
J. Needham
(Gonville and Caius, Sir William Dunn Reader in Biochemistry, F.R.S.)
Pethick-Lawrence
(Former Secretary of State for India, Sometime Fellow of Trinity College)
C. Winter
(Trinity, Director of the Fitzwilliam Museum)

All M.A.'s and holders of higher degrees are entitled to vote, whether they are resident or not. Voting, which is in person only, will be at the Senate House on Friday, 10 November 1950, from 9 a.m. to 9 p.m. Gowns will be required by the voters. Those who cannot bring them may obtain them in Cambridge.

Of all the people I interviewed in the first round, it was Joseph Needham who surprised me the most. He had been suggested to me by the biochemistry department as a likely supporter, and I contacted him immediately. He invited me to his college rooms for a prescribed 10 min. There was no need to discuss anything about the candidacy of Nehru or any other matters of the election. His first words were that there could be no question that this was the right choice. But then, I came to understand the reason for the clear time limit he had given me. He was very friendly, but he did not wish to interrupt his work for a moment. He got down books from his extensive library, opened pages, wrote down notes, and continued this without looking up. When I tried to inquire about this work, he stated briefly that he was in the middle of a multivolume work on the scientific history of China. This work is widely known now, published as *Science and Civilisation in China* by Cambridge University Press. I was glad to have seen the intense devotion that scholars can muster for their work. I left, with his signature, well before the prescribed 10 min. I came to know him quite well in later years, more relaxed, when he had finished this task.

Another person who came to join the team was my friend John Kendrew who later became a Nobel Laureate, but was at that time already widely known for his fundamental work for the modern subject of molecular biology. He had spent most of the war in India as the chief scientist to Mountbatten when Mountbatten was the head of the East Asia military command. It was during that time that preparations were underway for the freedom of India from British colonial rule. Nehru had fought for that freedom and had become the leading figure. The actual transition of power came later, guided by Mountbatten, who had been appointed the last viceroy of India.

Another person who expressed great enthusiasm for the election of Nehru was the writer of one of the great novels about India, E. M. Forster (*A Passage to India*), at that time, a fellow of King's College, Cambridge.

Joseph Needham urged me to go with him to London to meet Krishna Menon, the high commissioner for India and with that the representative of Nehru in London. Krishna Menon was a person who had great understanding for political situations and seemed to know exactly what the problems were in a campaign to elect Nehru to the chancellorship. He was obviously favorably inclined, but he also told us that he could not predict whether Nehru would be willing to accept that position or to be involved in a contest against Lord Tedder. We informed him that the rules for the election did not require any statement from the candidates concerning their willingness to serve, and we hoped of course that Lord Tedder would be equally unsure whether he should stand in opposition to the prime minister of India. If he withdrew, that of course would solve the problem immediately. The difficulty of making sure that he knew what was going on was that he was at that time in Washington, and we were not sure that he would be reading the British newspapers.

We started a large campaign sending out thousands of leaflets to the long list of M.A.s of Cambridge who were resident elsewhere. The staff of the Senate of course assisted us as they were obliged to do, after we had officially informed them of Nehru's candidacy, but we had to find ways of addressing thousands of envelopes to send out. Many helpers, undergraduates for the most part, came to our assistance. And Jack Gallagher had set aside his rooms in the college to accommodate all the envelope writers. In the middle of this, Jack Gallagher received a phone call from none other than Willinck, the chief supporter of Tedder. Willinck asked in the first place for me, but at that moment, Jack did not know where I was, and so Willinck spoke to Jack and asked him outright whether we had any knowledge whether Nehru would be willing to stand. Jack answered that of course we had no such knowledge.

When I got back to the operations room and Jack told me this, I said to him, "But did you not understand why Willinck would phone into the enemy camp? He knew the rules just as well as we did, that no such knowledge had to be obtained for this election. The only reason why he would have phoned would be that Tedder had proposed to decline to stand against Nehru and so Willinck wanted to get a negative reply from Nehru to give to Tedder." If Jack had only said that it was not required to obtain a candidate's consent, Tedder might well have withdrawn.

Willinck apparently let it be known in his circles that he did not think that Nehru would stand and I feel sure that Tedder was so informed and that if he held out, it would be an unopposed election. Nehru apparently withdrew when he got word that Tedder was willing to stand presumably communicated to him by somebody at Willinck's request. I received the following letter from Krishna Menon by the next post after this episode.

High Commissioner for India
India House
Aldwych, London, W.C.2.
4th November, 1950

Dear Dr. Gold

I am most grateful to you for the kind understanding shown by you and your colleagues to the request I have made on behalf of my Prime Minister. I know it must have been a very difficult decision for you to make, and if I may say so, you have done him a great honor in respecting his wishes with such understanding. I will not fail to convey to Pandit Nehru the assistance I have received from you, in spite of your own very strong views in the matter and all the other surrounding circumstances.

I much look forward to meeting you and other friends interested in India when I am in Cambridge next.

I have enclosed for your information *not* for publication, copy of my letter to the Vice-Chancellor.

Best Regards
Yours Sincerely,
Krishna Menon

Dr. T. Gold,
13, Barrow Road,
Cambridge.

I later found out that what I had suspected was precisely what had happened. Tedder had indeed phoned Willinck to tell him that he wanted to withdraw and not stand against the Prime Minister of India, and Willinck had informed him that it was most unlikely that Nehru would stand. So it was an unopposed election in favor of Lord Tedder. There is just one more item to this story and it is the following personal letter that Nehru wrote to me sometime later.

No.1905-PMH/58.

Prime Minister's House
New Delhi
August 8, 1958.

Dear Professor Gold,

Some little time ago, our Consul General in New York sent me your letter dated March 29, 1958 as well as the papers you had left with him. These papers deal with what is called the Nehru Election Campaign in the Cambridge University.

I wanted to read these papers before writing to you. I have now done so.

It was very good of you to send all these papers to me. I realise now how much trouble you took over this affair and feel a little guilty that I should have been the cause of all this trouble to you. I am convinced however that my decision not to stand for election was the proper one. Whether I won the election or lost it, it would have been embarrassing for me.

Again thanking you,

Yours sincerely,
Jawaharlal Nehru

Professor Thomas Gold,
Harvard College Observatory,
Cambridge 38,
Massachusetts,
U.S.A.

Spinning Dust in Space

In these 5 years in Cambridge after the war, I concerned myself with many different topics and wrote quite a number of papers. I suppose that was the purpose of giving out Prize Fellowships. The strange thing is, as I now look back on that period, I do not remember being very industrious at that time. In fact, I was rather lazy. Perhaps, the little poem I wrote at that time characterizes me:

The Worker's Morning
When dawn breaks then I face with vigor
another day's appalling rigor,
with so much turmoil, work and the strife,
and all the things one does in life;
I make my preparations then
by sleeping until half-past ten.
Then out of bed, up with the lark,
active like an electric spark,
I go down to the breakfast table
and eat as much as I am able;
and also read the Times and Mail,
the Chronicle, and without fail
the Daily Herald so I'll know
the way in which events will go;
on Wednesdays there's also Punch,
which makes it nearly time for lunch.
A rapid bath before I go,
(I only have an hour or so),
and then I'm off to go to work,
(a thing that I would never shirk),
though first I have to go and meet
some friends for lunch, for man must eat.

I certainly have been more industrious in later years when I had more demanding jobs, but probably it was in those carefree years that I developed the habit of thinking most of the time about one problem or another. Those might be "real" problems in physics or astronomy or geophysics, or they might be problems whose solutions were perfectly well known to some, but not to me. I did not like to study something from a book if I thought that I could puzzle it out for myself. It seems to me that this brings with it the advantage not only that one remembers it better but also that one understands better the connections between different fields, when one had to fish around for an explanation in one's head rather than find it presented on a platter.

One such problem gave me a particular insight into the way a brain works. Does it work only while one is concentrating on a problem, or does it work on that problem at other times, too?

I had been trying to find the explanation of the observed polarization of the light from many stars. Observations had made clear that this polarization did not arise from any effect at the stars but from some property of the interstellar medium that caused light of one polarization to be absorbed more than that of the orthogonal polarization. Elongated or needle-shaped grains of interstellar dust could clearly do this, if more of them were aligned in one direction than in another. Of course, if they were just jostling around, such as due to random impacts with gas molecules, the different orientations at any moment would be equally populated. The theory that had been proposed to account for an alignment of dust grains, due to Leverett Davis and Jesse Greenstein at Caltech, suggested that an interstellar magnetic field could be responsible, provided that the grains had the appropriate magnetic properties. I was not happy with that. The necessary magnetic field was stronger than I thought probable, and the magnetic properties of the grains were unusual for any kind of grains we thought would be there. It was not impossible that they were right, but it was certainly appropriate to think up other solutions to the puzzle. After actively thinking about that, I had not come up with anything.

One morning, about a week after I had given up on that problem, I awoke suddenly at 4 a.m., much earlier than usual for me, and it was some sudden shock that seemed to have awakened me. "Here is the answer to your problem" was the message. If the grains are in a stream of gas that possesses some velocity relative to them, they will not have random orientations at any one instant. Instead, grains spinning with their long axes in planes that contain the direction of the relative velocity will be favored over grains spinning with their long axes at right angles to that direction.

I thought of a rather silly example of this. Suppose there are soldiers in a wartime trench, and a fierce stream of machine gun bullets flies over their heads. They throw up their drum sticks above the trench, and some are hit. What would be the directions of their spins, and what would be the distributions of their orientations at any one instant?

Of course, this could easily be calculated. I switched on the light and wrote this down on a pad by my bedside, knowing that one tends to forget thoughts that occurred in the night. The next morning, I got up as usual and did not think of this at all. Later, I picked up the pad and saw the message to myself and of course remembered it all in a flash. I published two articles on this, giving some details of the degree of polarization of the light that could be expected from this effect. The criticism I got was mainly that there would be only a rather thin slice in which a relative velocity between gas and dust would be maintained. In a rather short distance, the dust would be picked up to the same velocity as the gas. It could not make as widespread a phenomenon as was seen.

Now, 40 years later, this problem has been reconsidered by Alex Lazarian (University of Wisconsin), who calculated the effect in detail for different shapes of particles and who added the important consideration that the understanding of magnetic interactions in the gas is important. Magnetically driven density waves exert their force only on the gas. They will, therefore, tend to regenerate relative velocities between gas and dust. Since the dust alignment caused is the same for one direction of the wind as for the opposite one, a magnetic density wave can cause alignment, but it will alternate the relative velocity to the dust so that the dust never comes to rest in the gas. Instead of the direction of a steady wind defining the alignment, it would now be the direction of propagation of the density waves. If such wave activity is as widespread as we generally think that it is, it will make large regions subject to this alignment effect of the dust grains. Quite possibly, this process will turn out the major one. If it is, the various deductions made for the magnetic alignment process will not apply. The grains will then not need to have the strange magnetic properties that had been assumed, nor will the deductions of the field strength or direction of the galactic magnetic field be valid.

But the main interest for me of this whole story was not whether it was the correct solution, but the light it shed on the workings of the brain. I take it that the sudden awakening with a solution was due to unconscious parts of the brain having continued to work on the problem from the time I had "instructed" them to do this. When they came up with the answer, it was their normal routine to knock at the door of that singular compartment which we call consciousness. There sits a doorkeeper who listens and who makes the decision what to admit. On this occasion, he might have said, go away, and try again when my master is awake. But in fact, he said, my master had given you such firm instructions to work on this, that I must admit you and wake him up.

But one can discuss all this in a more serious vein. The knowledge of the speed of working of the brain cells and their connections makes quite clear that a computer that works by taking all steps just in succession, a serial computer, could not do what our brains can, in fact, not by a very large margin. The computers, which are the objects you buy in the computer stores, have a very much faster action, and they can do quite a lot. But if you slowed them down by the factor by which our brain action and that of our nerves is slower than computer electronics, they would be useless. The only design of a brain that could possibly achieve the performance of ours is that of a massively parallel computer. (Let us leave out of discussion that our brain works by means that are new to physics, that the "life force" does it, and all such arguments that are frequently advanced with great passion but little knowledge and understanding.) Now a multiply parallel computer is very hard to design, and the modern supercomputer designers know this very well. A central system has to distribute the various parts of the problem to individual units, who must then present their answers again to a central unit. There they must be combined with all the other partial responses, and that combination then represents the information that is fed into the final single channel output.

The brain of an animal or a human must also combine outputs of parallel channels into a single final one. The animal has to respond with an action, and it cannot respond to a large number of different instructions simultaneously. This, I believe, is the channel we call consciousness. Whether all the subordinate channels also possess consciousness of their own, we do not know, just as we cannot look into the subjective consciousness of another person or another animal. In any case, I think that a grand plan of the design of a brain has to have the basic features I have described, quite irrespective of the detailed architecture involved. I can see no other way in which the performance can be achieved. The example with which I started out is not the basis for all this discussion; it was only an occurrence that reminded me of these considerations.

Chapter 5
Appointment at the Royal Greenwich Observatory

In 1952, when my Trinity Fellowship had run its course, a very nice gentleman, Sir Harold Spencer-Jones, the Astronomer Royal of Great Britain, offered me the job of his Chief Assistant at the Royal Greenwich Observatory. This was a great honor, after so short a career in astronomy, to be offered such a position by the most prominent astronomer in the UK. I think I would have preferred to stay in Cambridge, but, as they say, this was an offer I could not turn down. And so once again, I found myself working for the Admiralty, which was responsible for the Royal Observatory, but now I had the illustrious title Senior Principal Scientific Officer.

I moved with my wife and daughter to the south coast village of Herstmonceux where the Greenwich Observatory was now located. The Observatory had moved there from the original site in Greenwich near London, which had become unsuitable for optical astronomy. But the Greenwich meridian did not move, and you can stand astride it at the old observatory. The new central quarters of the observatory were in a magnificent fifteenth-century moated castle nestling in a parkland setting. My office was one of great splendor. Around it, on the castle grounds, were a number of buildings housing the various telescopes that had all come down from Greenwich. Some were 100-year-old instruments, still in good working condition and representing the best craftsmanship that had then been available.

My job was to become acquainted with the various departments and give advice where I thought fit. For any major changes, I would, of course, consult with the Astronomer Royal, who had his immense office below mine. He was always ready to talk with me, both about observatory matters and also about any researches in which I was engaged. There was no question that I could devote as much of my time as I wanted to my own research work, a very generous arrangement. I could draw on secretarial assistance whenever I required this, and also on technical assistance by skilled people from any of the departments. It was a very pleasant existence, though I lacked the stimulating conversations of Cambridge. But by train, I could travel frequently to London and Cambridge and did not really lose contact.

Thomas Gold *edited by* Simon Mitton, *Taking the Back off the Watch*,
Astrophysics and Space Science Library 381, DOI 10.1007/978-3-642-27588-3_5,

The department in which I took the greatest interest was the one concerned with observations of the Sun. Here I could watch the solar flares, the gigantic explosions that sent high-speed gas clouds to the Earth and far beyond, and also the majestic structures of the corona, the outer hot atmosphere, in which great arches formed and vanished again, where gases flowed at the strangest angles both upward and downward, and I could get a good view of magnetohydrodynamics at work. Disturbances of the Earth's magnetic field were also recorded by another department, and I could watch the correlations of changes in the magnetic field with the solar phenomena. When I later concerned myself in some detail with these matters, my education there was most helpful.

Most of the staff were people of very different outlook to my own. They regarded it as their business to collect information in the various fields and to publish it in a useful form. They were mostly highly skilled in their jobs, and they performed a very useful service to fundamental astronomy. Few of them had any interests beyond that, or if they did, they did not show it. They were very competent lifetime civil servants in the service of Her Majesty's Admiralty. I had considerable respect for them and got on well with them. But I still had to look outside that environment for the discussion of the meaning of the information they were archiving, or for the researches in which I was really interested. One such topic was the rotation of the Earth.

Axis of Rotation of the Earth

I had stumbled on this topic and its problems both through work at the observatory, for which this was a major topic, and through reading various texts. What is there to discuss? The Earth just turns once in 23 h and 56 min relative to the fixed background of the stars. Dynamical theory states that so long as there are no forces exerted on Earth, it will continue to do this. But oh! When you look at this in more detail, there are many problems. Firstly, the Earth is not exactly a sphere but is a little fatter around the equator. The gravitational force of the Sun does not only hold the Earth in its orbit, but it also exerts a slightly greater attraction on the nearer part of this equatorial bulge. As the rotation axis and the axis of the orbit around the Sun are at an angle to each other (about 23°), this force will exert a couple, and the spinning top that represents the Earth will then precess, meaning the rotational axis will move around, describing a cone. It takes a little less than 26,000 years to complete one cycle of this motion. It is a slow precession, but easily measurable. Famously, the Greek astronomer Hipparchus discovered the precession of the equinoxes in the second century BCE.

But precession is not the only change that occurs. There are some shifts of masses on the Earth that take place on the surface, such as atmospheric masses that are not quite equally distributed, and whose movements are related to the changes of atmospheric pressure we observe. They produce an extra load here and there, and the Earth will change its axis of rotation slightly in response to this. The apparent pole of rotation will move by a few feet from the position we regard as the pole (North or South). This is technically called a "free nutation," meaning a nodding not caused by any external forces. This free nutation can also be observed, but only with very precise instruments that look at the apparent change of position of a star, referred to the local direction of the force of gravity. Observations of this nutation showed that it has a certain tendency to return to the "no wobble" situation. This observation demands that there is a dissipative force at work that takes some energy out of the nutation and gets rid of it, so that the Earth does not continue to wobble forever as a result of each disturbance.

I read the discussion of this problem in the third edition of a book by Harold Jeffreys called "The Earth," a widely read major textbook of geophysics. The explanation Jeffreys gave for this "damping of the free nutation" presented a formula involving the quantities that enter into the problem, such as internal masses and viscosities of fluids. I stared at this formula for a while, and then I concluded that it implied that an object, however small, say the size of a pea, could cause the damping of this motion, if it had the right value of the viscosity that coupled it to its surroundings. Obviously, that was wrong. No limits to the applicability were given in the text.

I discussed this with Bondi and said mockingly, "Look here Hermann, Harold Jeffreys thinks that a pea could cause the damping of the free nutation of the Earth." Bondi laughed and then studied the few paragraphs in the book. "Obviously," he said, "this has to be corrected." The matter was of some importance since this was one of the few items of information we had of the mechanical properties of the interior of the Earth. We quickly wrote a joint paper and submitted it to the Royal Astronomical Society for publication, expecting that the correction of such an error would be accepted without delay. But in fact, it was not. We spent almost a year haggling with referees about the matter before the paper was accepted. It was the first time I had seen obvious failings of the review system at work. The joint paper was eventually published with the title "The damping of the free nutation of the Earth" (H. Bondi and T. Gold, Monthly Notices of the Royal Astronomical Society, Vol. 115, No. 1, 1955). In this we describe the error made by Jeffreys, essentially a confusion that led to an error which, when expressed in numerical terms, amounts to a factor of 1.8×10^5. We demonstrate that whatever value the viscous coupling between the solid mantle and the liquid core may be, the core is just too small to be responsible for the damping of that motion.

I wonder how many students of geophysics had read Jeffreys's book and not noticed an error of so large a numerical value! To me, this episode illustrated how important it is in science to carry along a good notion of the magnitudes of the quantities that are involved in a problem. I remember the bad old days of the slide rule, before the calculators. Then it was quick and easy to get an approximate result to a calculation, but one always had to keep tabs on the powers of 10. I found it best to have a good guess handy of the expected result. I never trusted an elaborate calculation without that.

Having concerned myself with the nonrigid behavior of the Earth, and having found this to be due to the plasticity of the mantle, it seemed natural to think of the greater problem: would the rotation axis of the Earth be stable over long periods or could it move? Of course, the axis of the angular momentum had to stay fixed relative to the stars (except for the precession forced by the Sun which I discussed above), but that by itself does not mean that the Earth must remain fixed relative to this axis. Conservation of angular momentum does not imply constancy of the orientation of a nonrigid body. A cat that is dropped upside down without any angular momentum has no trouble turning over to land on its feet. (You may wish to think out how the cat does this, though you may find it unkind to try it out with your cat, but I promise that it will work. The cat has a profound understanding of nonrigid body dynamics.) I asked myself, could the Earth in fact have tipped over by a large angle in geologic times?

If the Earth were a perfect sphere, it would have no permanent stability, and the smallest change of mass distribution would cause it to tip by a large angle relative to its axis of spin. The poles would move relative to the surface. What stability it has in fact is due to its nonspherical shape, namely, the bulging out of the equator. This is of course interpreted as the consequence of an adjustment of the shape to the equilibrium defined by gravity and the centrifugal force. For a rigid body, this would imply permanent stability of the position of the poles (since spinning around the axis of the largest moment of inertia represents the minimum spin energy for a given value of the angular momentum). But then, a permanently rigid Earth would not have made the quite precise adjustment of its shape to the equilibrium defined by gravity and spin.

Sir George Darwin (the second son of Charles and a major figure in geophysics) had worked on this problem and together with Lord Kelvin had come to the conclusion that there would indeed be long-term stability, even though they had recognized the plasticity of the mantle indicated by its adjustment to the equilibrium figure. A small shift of masses, such as those indicated many times in the geologic record, would not make a large angle change in the axis of rotation relative to the body. The Earth would not have tipped over, and the climatic record of geology should therefore be read with reference to the present position of the poles.

Also, Darwin thought that his conclusion, and that of Lord Kelvin, was confirmed by the absence of evidence in the geologic record for large changes in the distribution of land and water, which in his view would have taken place with any large tipping of the axis. Of course, in his day, the facts were not yet available, that the high and low areas on the Earth were largely the result of crustal blocks of greater and smaller density, all floating in equilibrium on a plastic mantle, like ships of differing heights all floating in a common sea. Gravity measurements had later established this effect, called "isostasy." If Darwin had known that, he would have understood that his argument about changes of the distribution of land and water was invalid, and he would not have given this as supporting evidence for his theoretical conclusion.

I came to the opposite conclusion and considered their arguments to be in error. Of course, it must be said that I had some incentive they did not have, namely, that paleomagnetic (fossil magnetism) measurements had indicated major changes over geologic time of the direction of the Earth's magnetic field, some of which could be accounted for by corresponding changes in the position of the spin axis of the Earth.

My conclusion, that large changes were possible and even likely, was based on the following considerations. Suppose the Earth was spinning freely, with its axis of figure adjusted to its axis of spin. Suppose that then a small extra mass (or a small extra elevation of an existing mass) was added somewhere in the midlatitudes. The major moment of inertia would now be slightly out of alignment with the spin, and a damped nutation would result. The axis of spin would be brought to coincide with the major moment of inertia, and this would be achieved by a small change in the orientation of the spin axis relative to the Earth (insofar as the disturbing effect was small). But this is not where the matter ends, as, I believe, had been assumed by the earlier investigators of the problem. Now the spinning Earth will still see its equatorial extension slightly displaced from the equilibrium shape, and in the course of time, at a speed determined by its plastic flow time constants, it will adjust its shape to the new, slightly displaced spin axis. But that has not cured the problem. By this, the axis of the principal moment of inertia has again been changed, and the body is again, much as before, spinning slightly skew to it. These adjustments will have to repeat, and there will be no final equilibrium reached until the equator has moved around, and with it the axis of figure, to include the disturbing mass in this final equatorial plane.

I have described the process as if it went in steps, but in reality it will be a continuous slow movement. A sudden impact with an external body, such as an asteroid, can of course make instantaneous changes.

What all this means is that a small disturbing force which would initially have produced only a small wobble and displacement of the spin axis will eventually result in a large change of the direction of the spin axis relative to the body of the Earth. I must emphasize again that in all this the axis of the angular momentum has remained strictly fixed. In the article I published in *Nature* to explain all these complicated matters, I referred to the donkey who runs to catch the carrot dangling in front of its nose, not realizing that it is dangling from a stick held by the rider. Running fast for the small distance the carrot is in front of him will not gain him access to it. In our case, the donkey does get the carrot in the end but not before having run a much larger distance than he had anticipated. A small disturbing force, such as the building of mountains in mid-latitudes, will have the effect of a slow but eventually a very large tipping of the spin axis.

Of course, I had to feel very secure with my deductions before submitting them to *Nature* since I was contradicting two giants, George Darwin and Lord Kelvin. Professor Walter Munk had also thought about these problems, and he submitted a paper to *Nature* soon after mine that supported me in this and that gave the history of this problem. There had apparently been no work published on this topic between George Darwin's paper and mine.

If there have been large shifts of continental plates, then the mass distributions or their heights (both enter into the determination of the principal moment of inertia) would have changed significantly to cause large movements of the poles. The records of the ancient magnetization of the rocks and the ancient climatic records must be interpreted with this in mind. Large movements of the position of the poles must be expected, if major movements of continental plates have occurred. Therefore, in the interpretations of the paleomagnetic and climatic records, an explanation in terms of the movement of the poles must be attempted first since this is almost inevitable if continental plates have changed their positions, but movements of the poles could also be caused by other internal effects of which we may be unaware. It seems that this view is now well established among those who study the dynamics of the Earth.

Dr. Joseph Kirschvink and colleagues at Caltech and Dr. Bernard Steinberger and colleagues at Harvard University independently established that brief periods in geologic history show that large-distance polar wander occurred in comparatively short periods of the geologic record.

The Surface of the Moon

The next subject that attracted my interest, as I was sitting in the isolated splendor of my quiet office in the Tudor castle, was the Moon. What had formed its visible features? What dynamical evolution accounted for its orbit? How was it formed in the first place?

These were all questions that had been mentioned in the literature, long before the beginning of space research with modern means, and some guesses had been advanced. Mostly the Moon was thought to be an intensely volcanic object, and its large flat plains were thought to be oceans of frozen lava. The multitude of craters were considered to be volcanoes. The scale of volcanic activity was thought to have been enormously greater than on Earth.

I had some doubts about this. Why did the craters have such large diameters? Could internally caused eruptions result in circular features of as much as 1,000-km diameter? Could lava have flowed out over hundreds of kilometers and remained liquid before freezing? Could the Moon have been so intensely volcanic in its past and yet have no signs of active volcanism at present? The telescopes on Earth at the time could have seen even quite minor volcanic eruptions. An active volcano on the Moon as intense as any volcano on Earth among the first 700 in intensity would readily have been recognized, yet no such observation is on record.

In 1949, Chicago University Press published a small format book by Ralph Baldwin, entitled *The Face of the Moon*. In it he discusses some of those issues: are the craters predominantly caused by eruptions or by the explosions resulting from high-velocity impacts? Is the Moon a giant volcanic structure or a cold body heavily damaged by external impacts? He made many detailed comparisons with the corresponding terrestrial features and came down heavily on the side of externally caused effects. I found this book in the early 1950s, and it accorded so well with my opinions that I decided to investigate the subject in more detail.

It became clear to me that one could decide quite definitely the age sequence of some features. A circular crater feature must be younger than a feature representing a part of a circle that is interrupted by another complete circle. No way could be seen whereby a part of an accurate circle would be created just adjacent to a complete circle. On the other hand, the creation of a complete circle would have overpowered a previous complete circle and left behind only a C-shaped crater wall adjoining the new circular one.

There were many examples of such features for which the time sequence could be established. Multiple overlaps allowed in some cases as many as five craters to be placed in a time sequence. That by itself would perhaps not have taken us much further, were it not for the clear observation that in all those cases the apparently older craters were less high and more rounded in profile than the younger ones. Possibly, in a few cases, this distinction was arguable, but in any case, it was never seen in reverse. But then, there were many cases in which the difference was very large: there were craters whose rim was barely perceptible while others in that same sequence and of similar diameter were steep and at least ten times as high. The passage of time must have generally caused a lowering of the crater walls and a smoothing of their profiles. What could that process be? On Earth, one would have said that this was due to erosion. The continuous action of wind and water would have ground down the older features more than the younger ones. There would be nothing surprising in that. But on the Moon? There is no wind and no water, and nothing we knew that could erode the surface. This was clearly an interesting puzzle.

Another observation supported the explanation in terms of an erosion type of process: several of the large, steep-walled, and apparently youngest craters had streaks of a different color radiating out, sometimes over dimensions comparable with the size of the entire Moon. If one did not want to suppose that these were made by a process different from that which made older craters of similar diameters, then one had to suppose that such a pattern of streaks had been produced in every impact of that size but eradicated in the course of time, and in fact in a time short compared with all the other time sequences, since only a small number of the very youngest craters had these features. These streaks must therefore have been removed or covered over by a process that continued to the last phases of the evolution of the lunar surface, and probably to the present day.

The streaks were presumably due to an ejected spray at the time of the impact explosion, certainly not due to cracks that went across the entire Moon, as some investigators had thought. Such spray may well have a different color from that of the rest of the surface, even if it were derived from the same material, since it will have had a drastically different treatment in the explosion vapors, and a shorter exposure history on the surface. In general, the streaks were lighter, and as we now know, surface exposure in the particle bombardment from the Sun will in fact cause a strong darkening of rock powders. The streaks clearly corroborated the age sequencing we had discussed, connecting overlaps and sharpness of features. But then to the next problem: how could material have been removed gradually from the old craters to make them much less high than young ones of the same diameter? In what form was this material that could be moved without wind and water? And where is all this material now?

The "missing" material is quite a large quantity, as judged by the total amount of "missing" height of all the old craters. A rough estimate would say that it could make a blanket of more than 1-km depth all over the Moon. It is unlikely that it was all propelled to escape velocity and thus left the Moon altogether. Small impacts would be needed to reduce the heights without destroying the features completely, and small impacts do not impart high velocities to the material that is struck. But then, why do we not see that material?

The answer I suggested then, and of which I am quite certain now, is that the flat plains that are so common both inside older craters and in all the low-lying terrains represent in effect that material. That view is, of course, in sharp conflict with the widely held view that huge lava floods had flowed over much of the Moon and that these "seas," as Galileo had called them, were the frozen lavas. To account for all the missing material, many of these flat deposits had to be about 2 km deep.

When it was thought that the surface attested to immensely intense volcanism, the lava flooding theories seemed an obvious conclusion to many. But I could not see that a body that possessed extremely violent volcanism in its past would then have cooled down to the extent of becoming an extremely quiet one. When it had become clear that the craters were mostly or all due to impacts, the case for intense volcanism seemed to me to have vanished.

If the maria were indeed deep deposits of the material that had previously been in the crater walls, what could have caused it to move? In what form would it have been to have become so mobile? What forces were there on the lunar surface that could be held responsible?

Here any analogy with terrestrial erosion processes is of little value. In the vacuum of the Moon and in the intense bombardment by radiation and high-velocity particles, quite different forces could come into play. But it is hard to imagine that anything other than high-velocity meteorite impacts could move large rocks or even pebbles, yet the flat plains seen as a repository of the missing material could not be made so flat by large impacts, without the appropriate impact craters being visible now.

My conclusion was that the material was transported as very small grains that could be moved by electrical effects, expected to be present in the lunar environment. We all know how dust in the house is often attracted to plastic surfaces, clearly by static electricity. I speculated how the bombardment by electrons and protons from the Sun, to which the Moon is subjected, would provide the motive power in the vacuum of the lunar surface to lift and agitate small grains and thus allow them to move downhill without friction. Of course, this would mean that the mountains would have to be predominantly made of rock dust so that much of their material could be eroded away, and the plains almost exclusively of that, except for rocks that would have frozen from the melt created in major impacts and that were later shot around by other impacts. What I did at this time was to publish a long paper explaining my conclusions. At that time, my views were generally appreciated. Radar results first obtained in Britain strongly confirmed them by showing the surface to be remarkably smooth on the scale of a few centimeters, yet impacts in frozen lava would have made very rough surfaces. Later, when a large number of people were drawn into the subject by the space program, there were many who wanted to preserve what they could of the previous views, irrespective of all the observational data. We will return to this topic later, in the context of the US and Soviet space programs.

Eclipse Expedition

In 1953, Spencer-Jones asked me to look into the possibility of setting up a telescope in the path of the upcoming solar eclipse of June 30, 1954, for the purpose of obtaining another verification of Einstein's predicted bending of light near a massive object, the Sun. Such an observation could be carried out only at the time of a total solar eclipse, at which the sky around the Sun could be dark enough to photograph the stars in the background. One could then, 6 months later, photograph the same star field when the Sun was out of the way and determine whether the apparent positions had changed by the small amount that the Einstein distortion of space would have predicted. The first such observation had been carried out shortly after Einstein's publication of his General Theory of Relativity, under the guidance of the theoretical astronomer Eddington. This had given a positive result, but perhaps not far above the error limit, and it would certainly be desirable to repeat such an observation.

Of course, I was delighted with this assignment and soon traveled to Sweden to the eclipse path, to select a site at which the quality of viewing and the local weather in June would be judged optimal. I found such a location on a little island called Syd Koster, off the west coast of southern Sweden. There I selected a clearing in the woods, about a mile from the nearest village and quite close to the sea.

The next job was to select and arrange the equipment. The Royal Observatory owned an antique but excellent refracting telescope of focal length 21 ft, well suited to this work. Of course, I would have to arrange for a structure of a corresponding size to house it. The telescope was taken apart and crated, and I designed a very large tent, some 30 ft long and about 20 ft high, constructed of steel tube scaffolding and canvas sheets. After some experimentation, this was all shipped off to Syd Koster. How we got all this on a little path through the woods I cannot remember anymore. (I am reminded of the cartoon showing the great structure of Stonehenge, with two little Druids standing in front of it, one wiping his hands and saying to the other, "Well I don't know how we did it, but we did it.") My helpers in this undertaking were Alan Hunter, a senior astronomer at the Royal Observatory, and two volunteers, Edward (Ted) Bastin, a young student of philosophy in Cambridge, and Dennis Sciama, my friend of many years. We spent 2 months building up the structures without any outside help. Living conditions were somewhat rough. The daily bath was in the freezing cold seawater. Food was bought in the village and cooked at the site; sleeping quarters were sleeping bags in the tent; water was carried in two buckets hanging from the handlebar of a bicycle until I decided to turn into a water diviner and instructed my crew to dig a hole just outside the tent.

I had noticed that there was a long sloping face of granite facing the tent and that this ended in our vicinity in a flat plain of sand. We dug the hole, and at a depth of not more than 5 ft, we found plenty of clean water that we could just lift out with a bucket.

The main enemy we had to face were the mosquitoes, so thick in the woods that it looked like a layer of fog. This was before the invention of adequate insect repellents, and so any work outside the tent caused exposed skin to be densely covered with bites in a matter of minutes.

Despite discussing deep philosophical matters with Bastin most of the time, we got the job done. The morning of June 30 was clear, the telescope set, a drill of changing plates in a few seconds had been practiced ad nauseam, and we were anxiously waiting. When the time came and the eclipse was just beginning, a little faint cloud came along and just plunked itself over the Sun and its immediate surrounds. What could we do? Of course, we went through our procedure and took the various exposures nevertheless, but it was already clear that the star field would not be seen. That one little cloud had ruined the work of months. While indeed no stars could be seen on the plates, the cloud had been sufficiently thin to show the corona of the Sun through it, the consolation prize of eclipse observers. It turned out to be quite an interesting picture, having been taken at a time of minimum solar disturbance, and therefore showing great regularity and telling us what the undisturbed magnetic field of the Sun looks like. We packed up the equipment and returned home, somewhat disappointed but not too surprised, for we had known that there would be only a 50–50 chance of cooperation by the weather.

I had settled very comfortably into the Herstmonceaux scene. I had a house quite close to the castle, which had been coach houses and stables of a large estate in a beautiful setting with lawns, trees, and a pond. With my own modifications, it had become a very comfortable modern home. My work was enjoyable, and I might have been turned gradually into a regular British civil servant. But luck would have it otherwise.

Appointment of R. R. Woolley as Astronomer Royal

By the end of 1955, Sir Harold Spencer-Jones was retiring, and a new Astronomer Royal, the eleventh, had to be appointed. It is an appointment by the Crown, but on the advice of a committee of distinguished astronomers, or so it is said. (Many years later, information about the deliberations was leaked to me.) The new appointee was one Richard van der Reit Woolley, the Commonwealth Astronomer in Australia, with responsibility for the Mount Stromlo Observatory in the Australian federal capital, Canberra. I knew nothing about him but was told that he was "a man of action" (in contrast to a man of thought, I wondered).

When he arrived at a British port, he was interviewed by the press. The statement that appeared in many papers was "New Astronomer Royal says space research is utter bilge." I had at that time been one of the astronomers urging the launching of a small instrumented satellite, which was clearly becoming feasible. With Woolley being so against space research, a position he first adopted in 1936, there was obviously some prospect of conflict between us.

Woolley arrived at the Castle and soon wanted to hear what the plans were that we had made for the future. The first item I mentioned was a project of mine, an addition to the solar observatory that had just been completed. It was a ground-based instrument that is sensitive to corpuscular radiation from the Sun, which, as we had come to understand, was strongly correlated with visible phenomena on the Sun. Since ours was the main solar observatory in Britain, and since this was a further line of investigation of newly discovered solar phenomena, it had seemed to me an obvious step toward the modernization of that department. A special small building had been constructed for it on the grounds of the castle, most carefully designed to minimize disturbing effects. The instrument itself was a large assembly of nuclear counters, at that time by far the most sensitive and best instrument of this kind in the world. I knew that in Australia, Woolley had run a solar observatory along classical lines, observing sunspots and solar flares, for example. Woolley responded to our imaginative plan to observe atomic particles from the Sun, without hesitation or argument: "Well, you will have to get rid of this. This has nothing to do with astronomy."

So far, so bad.

My next item was my plan to expand into radio astronomy. The observatory did a lot of work in astrometry, the precise measurement of stellar positions. It did this with beautiful, somewhat antiquated instruments called transit circles and zenith tubes. One of the purposes was, in fact, the observation of the timekeeping and the wobbles of the Earth which I mentioned earlier. All of that harked back to King Charles II who in 1675 had appointed John Flamsteed the first Astronomer Royal, who was "... forthwith to apply himself with the most exact care and diligence to the rectifying the tables of the motions of the heavens, and the places of the fixed stars, so as to find out the so-much desired longitude of places for the perfecting the art of navigation."

What I was now proposing was not inconsistent with the spirit of that royal command. Indeed, I had had the agreement of Spencer-Jones, who like Woolley had been against space research. My plan was to see whether in the course of time we could perhaps develop radio instruments that could give an even higher precision of positions of stars and with that of the small wobbles of the rotation of the Earth. Why did I think this possible? Because I had thought that the techniques developed by the pioneers Wilbur (Chris) Christiansen in Sydney and Martin Ryle could probably be extended to make interferometers with a spacing comparable with the entire size of the Earth. The shortest wavelength that could be used, I guessed, would be a few centimeters, the shortest waves that would not be messed up by atmospheric disturbances. Since the angular resolution of an interferometer is inversely proportional to the number of wavelengths that go into the baseline of the instrument, this would represent a far finer definition than any optical instrument. It would require the development of a technique to transport a signal from one end of the Earth to the other and maintain the phase precision needed for such an interferometer. One could record the signal in each location together with the signals from a very accurate electronic clock and then combine them at leisure later. Woolley did not listen to any of this at all. He responded with "You can't even discern one degree with a radio instrument. How do you expect me to take seriously that you could measure to one hundredth of an arc second? In any case radio does not belong in astronomy."

So much for that. By now, the technique is in standard use and goes under the name of very long baseline interferometry (VLBI), and it gives indeed the highest-resolution angular measurements in astronomy.

I withdrew from further discussions of this kind with Woolley. Evidently, we were not on the same wavelength. I had to start thinking about finding myself another job.

It was a remarkable coincidence that just a short time after Woolley assumed his Crown appointment and told me to throw out the brand-new solar monitor, the Sun produced the biggest display that had ever been recorded. My instrument, not yet thrown out, was the best for documenting the particle radiation of this remarkable event. Many papers were written by us and by others about it, and our records were the most prominent item. But this did not change Woolley's mind, and the instrument was dismantled. "This was not astronomy, and the Royal Observatory had no business to concern itself with such topics."

Another occurrence further confirmed to me that I had to leave the Observatory. I was responsible for arranging a weekly colloquium for the staff and occasional visitors. On one occasion, I had selected a young member of the Observatory staff to give a presentation of some very nice work he had done in celestial mechanics. Woolley had indicated to me that he would be interested.

These colloquia took place in the former chapel of the Castle. Woolley's palatial chamber was right next to this small meeting room. The room was full, and the starting time had come around, but as Woolley was not present, I suggested out of respect for his position that we delay by a few minutes. The young speaker was visibly nervous, and this did not help him. After 15 min, I decided that we had to start. The speaker duly read from a carefully prepared text which even included some humorous remarks. When 10 min of this had gone by, some 25 min after the supposed starting time, the door was flung open loudly and Woolley strutted in, walked to the front row, and threw himself into a chair. Then he bellowed out: "Start again, will you." The poor speaker then had to read his text again, including, with great embarrassment, the humorous passages. I do not know why I did not walk out of the room at that moment.

Chapter 6
Move to America

Palomar Observatory and the Crab Nebula

In 1955, I had attended a cosmic ray conference in Mexico, and I took a swing through the United States on the way back. My first stop in the USA was at the Mount Wilson and Palomar Observatories Pasadena where I was to meet again with Walter Baade, the highly successful astronomer and observer at the 200-in. telescope. The year before, I had discussed the problem of the Crab Nebula with him. This patch of milky nebulosity in our galaxy had been an interesting enigma to astronomers for some time, since the origin of the light from it could not be identified. There were some stars within the patch, but none sufficiently bright to cause a total light emission from the nebula that was judged to be about 100,000 times the output of the Sun. The Dutch astronomers Jan Oort and Theodore Walraven had just published (1956) a surprising new observation, namely, that the light from this nebula was strongly polarized. They gave a figure of approximately 6% polarization, which was high compared with other astronomical objects. They used a novel photometer attached to the 13-in. photographic telescope at Leiden Observatory, and they had been able to take only the light from the entire nebula for this observation, not from any specific part. I told Baade that I could not understand any mechanism that would make an overall polarization, but I could understand that perhaps the radiation came from fast electrons spiraling in magnetic fields. The theory of this effect (known as synchrotron radiation, first seen in particle accelerators) had been worked out initially by Hannes Alfven and Nicolai Herlofson in Norway. It seemed to me that it could apply to this case.

Since the magnetic field of such an unkempt-looking nebula could not be expected to be all in one direction, it would be necessary for portions to show a much stronger polarization so that the average over the whole object might be the observed 6%. Why not photograph it through a polaroid filter in various orientations and see whether the picture then breaks up into separate regions with much stronger polarization? Of course, this was a somewhat bizarre suggestion when people were wondering even how one could get as much as 6% and when no optical radiation

Thomas Gold *edited by* Simon Mitton, *Taking the Back off the Watch*,
Astrophysics and Space Science Library 381, DOI 10.1007/978-3-642-27588-3_6,

from fast electrons had yet been seen anywhere in the sky. But Baade took very kindly to it and promised to make the observation at Palomar, where there would be ample brightness available for a more detailed examination of this feature.

So when I arrived in Pasadena, the headquarters of the Palomar observatory, Baade greeted me with the good news that he had a span of three observing nights at Palomar. It was just the best time of the year for observing the Crab Nebula, and I should drive up with him. I was delighted.

We arrived midday on the top of the mountain, and I was given the royal tour of the facility. (In my opinion, even 40 years later, it still is a breathtaking piece of engineering!) Baade had brought up a simple rotatable polarizing filter, and he was going to place it just in front of the photographic plate at the prime focus and take at least three exposures with 45° rotation between them. The prime focus of the telescope is in a little cage high above the largest piece of glass I had ever seen, the parabolic mirror of 200-in. diameter. As the nighttime temperature on the mountain was quite cool, Baade dressed up in his electrically heated suit and, defying the no smoking regulation, lit his Havana cigar. I stayed up there for a little while and went down when I got cold. I had the intercom and chatted with him as he was taking the plates. When finally he had finished, it was early morning, and we went to bed just for 3 h. Then a quick breakfast and we were off to the darkroom to develop the plates. When they were done but still dripping, we hung them up in front of the light, eager to see what we had got. It was a very exciting surprise: each of the three pictures showed a completely different shape of the nebula. (A fourth would just represent a 180° change, which is the same as zero for polarization.) Not one picture looked like any other, or like the pictures that had been taken earlier, without the polarizing filter. Clearly, different regions were differently polarized, and in each, the polarization was very strong, near 100%. Nothing like this had ever been seen before. We stared at these pictures for a long time but could not really make out what they meant. I suppose we were too sleepy by then.

Baade returned to his high perch in the telescope again the next night for some other observations. Since I had nothing to do with that, I returned to the darkroom to study the photographs again. The plates were dry now, and this allowed me to handle them and to make better comparisons between them. Baade had also taken a picture without the polarizing filter, and this allowed me to superimpose the filtered pictures with this one. Suddenly, the answer hit me; it was so clear that I was quite embarrassed not to have seen it before. Wherever the unpolarized picture showed an elongated filament, the picture with the polarization direction nearest to a right angle to the elongation of that filament also displayed it, while the filaments that were aligned parallel with the polarization direction could hardly be seen on the polarized picture. I tried this out with all three plates, comparing each one with the unpolarized one, and the rule held every time. There was no doubt left in my mind about the meaning of that: the filaments were elongated in the direction of the local magnetic field, different in different parts of the nebula, and in each case electrons were spiraling around the field lines. Their optical synchrotron radiation would then have to be polarized at right angles to the local field.

With great excitement, I told Baade this over the intercom. It was an effect of remarkable implications, for it meant that we were looking at an object 100,000 times as bright as the Sun, and most of its brightness came from very energetic electrons. This huge amount of energy had come from some particle accelerator that must be sitting somewhere inside the nebula, but that did not itself make much light. We had glimpsed the first celestial light source seen that was not powered just by the heat of a star. It was also the first time we had evidence of powerful celestial particle accelerators and of light being produced by them. It was the beginning of the subject that is now called "high-energy astrophysics," which has become a major branch of astronomy.

When Baade came down from the telescope and into the darkroom, he stared at the pictures with great astonishment. "How was it possible that we did not see the effect yesterday?" He kept finding more patches where the rule was satisfied, sometimes so completely that almost 100% polarization was implied. I had another encounter with the Crab Nebula more than 10 years later, in the discussions that followed the discovery of the pulsars. It turned out that the faint star that Baade had always regarded as the source of the light, though by unknown means, was indeed the pulsar and, with that, the giant particle accelerator. I shall return to this aspect in Chap. 8.

Baade and I returned to Pasadena in a state of high excitement. Our discovery immediately became the talk of the astronomers and the physicists at Caltech. Baade gave me copies of the pictures and allowed me to show them on my further travels through the States, even before their publication. I did this with great delight and, of course, made careful reference to Baade. I remember in particular the discussions after my colloquium in the physics departments of Chicago, at Harvard, and at the Massachusetts Institute of Technology.

This tour of the various centers in the USA turned out to have had great implications for my future. After my return to England, I received several offers of professorships from American universities I had visited. It was at a point in my career when I was sure I had to look around for another job, and I knew that no professorship in astronomy was available in Britain. So I was very glad to receive these offers. The invitation from Harvard University was of the greatest interest to me. I had earlier responded to an invitation from Cornell University and agreed to spend one semester there, just to see how well I would like it there. I had liked Cornell when I visited there briefly on my US tour, at the invitation of Arthur Kantrowitz, to give a colloquium on the subject of my theory of hearing. I think it was the first time I presented this to an audience who understood what I was talking about. Among the people at Cornell whom I found attractive, there were Kantrowitz whom I knew from the Cambridge meeting on gas dynamics and from his subsequent 3-month stay in Cambridge, Phil Morrison whom I had met at the Mexico conference, and Ed Salpeter whom I knew from an earlier visit he had made to Cambridge. But I also liked the countryside and many other people I had met there on my brief visit.

The way in which the Harvard invitation reached me was amusing. It was normal practice in Britain at the time for the post office to phone to inform one that a telegram was waiting and, if so requested, to read it out before it could be delivered. So the operator at the local post office in the small village of Herstmonceux phoned me. I am sure she knew who I was, but we had not met. She said, "I have a telegram here for you. It is from America. I know you will like this. They are offering you a professorship with a salary of so-and-so much" (a figure that must have sounded astronomical to her). I asked her to read it out to me; it was from Donald Menzel, the director of the Harvard College Observatory. It was a particularly kindly worded invitation, and it spelled out the details of the proposed appointment. I would be one of the youngest Harvard full professors, I would be in charge of the Harvard radio astronomy section, but otherwise I would be largely free to choose my work.

I decided to accept this offer, and my wife, who was American, was pleased with this decision. My reply stated that I was already committed to go to Cornell for one semester but would then accept the Harvard invitation.

My father had died, and my mother was moving back to Vienna, where she would feel more at home when she now had to live alone. Although I liked living in England, I had to tear myself away. So here, in the summer of 1956, we sold our house, much of which I had built myself, and most belongings, and flew off to the States. In the first place, we were committed to go to Ithaca, New York, for my one-semester stay at Cornell. It was a very enjoyable time except for two problems: one was that we had rented a lovely house in the country, but an early winter hit us, and traveling to and from work proved very difficult. There were many occasions when I had to crawl under the car in really freezing temperatures to put on snow chains when I had got stuck. The other, and more serious problem, was that I contracted a rather rare but serious disease called recurrent pancreatitis. It is a disease akin to hepatitis, but one that has its ups and downs over a few years. Possibly, I got it from raw oysters, which I loved and which I had never been able to afford in Britain. I had several hospital investigations, but there was no known cure.

In midwinter, we moved to Harvard and bought a house in the country, in Lincoln, in the outskirts of the Boston–Cambridge area. It was a beautiful location, but it required a lot of driving to work every day. Houses nearer to the Harvard campus existed but were at a price that was unimaginable to me at the time. In any case, I did not like living in city surroundings, and even then, there were already gangs of youths roaming the city streets at night.

Work at the Harvard Observatory was very pleasant and interesting, and I liked it, as well as the freedom of choice of my work I had enjoyed in most of my previous scientific career. American university life was still unfamiliar to me. I was quite ignorant of all the procedures of running a department, applying for research funds from government agencies, but also of the undergraduate life and its problems, when I was expected to advise and assist. My research work was well appreciated; all the members and staff of the observatory were very friendly, and I enjoyed their company. At nearby M.I.T., there was Bruno Rossi, the leading cosmic ray physicist, and his excellent group, now gearing up to expand their activities beyond the field of cosmic rays to high-energy phenomena and soon to space physics. My Crab Nebula story was a major item for a while and so were some of my researches into solar system magnetohydrodynamics. Rossi invited me to give a course of lectures at M.I.T. on these topics.

Another enjoyable aspect was my contact with the Harvard Physics Department, and in particular with Edward Purcell, known for his role in the discovery of the radio emission at 21-cm wavelength of atomic hydrogen in space, which soon became a major tool in astronomy, but known also for the prediction and observation of nuclear magnetic resonance. Today, one hears of NMR as an important, noninvasive tool for diagnosis in medicine, but in the first place, it was a revolutionary tool in nuclear physics. Purcell had received the Nobel Prize for this, and even in the highest intellectual circles at Harvard, he was nicknamed "the brain." I got to know him well, and he was my best lunchtime companion and the best person with whom to discuss new ideas in almost any field of science.

Some of my lunches I took at the Harvard faculty club, which gave me the opportunity to meet many other members of the Harvard faculty. I remember one little amusing episode there. I was sitting at a table with several others whom I did not know. Two of them had a conversation about the anatomy of an owl, in particular the barn owl. I was quiet in order to listen to conversations at the same table, and I heard them discuss the oddity that the barn owl had unsymmetrical outer ears. One ear had the ear canal in the upper half of its external ear, and the other in the lower. While most animals are more or less symmetrical, this feature was not a singular aberration but consistent through the species. Why? Was there some advantage gained by that? At this point, I chirped in, "May I explain?" "Yes, of course." "With symmetrical ears there has to be the median plane of symmetry and no auditory direction finding is therefore possible within that plane. Whatever noise the owl heard that originated in that plane, both ears would receive the same signal. If the two outer ears were different, there would still be a plane of ambiguity, now at some angle and no longer the medial plane. That would seem to have gained nothing. But now, as different frequencies of the sound will be treated differently by the two ears, the ambiguity plane will be a different one for each frequency. If the owl hears a noise containing several frequency components, and can assume that they all came from the same point, then it can resolve the direction in three dimensions in most cases. A rustling noise in leaves would be such a signal."

"Splendid," was the response, "are you a biologist?"

"No," I said. "I am an astronomer." I then had to admit that I had been asked just that question before, namely, by my colleague Pumphrey in Cambridge, and I had devised the answer then. It was just my good fortune to come into this conversation.

The sound direction-finding requirements of the owl are different from those of most other animals. The owl needs to know the direction to a source accurately and quickly. There is no time to turn the head or move an external ear and observe the resulting changes of the sound in the two ears. Also, the owl in flight must know the angle to the source in both the horizontal and the vertical plane to good accuracy. The unsymmetrical ears give it that facility. Ear movement for direction finding, as many land animals use, is impractical for the owl, since it has to depend on extreme perfection in the aerodynamic design to avoid airflow noises that would mask the noise it wants to hear. My colleague Pumphrey had in fact followed up my explanations with detailed experiments using a barn owl in a barn and observing it diving for a bait that he had equipped with an adjustable noise generator. He could then observe that the wide frequency band was indeed a requirement for the owl to dive accurately, and he observed several other detailed consequences of this interpretation.

The designers of modern telephones would have done well to investigate similar problems before changing from the old and familiar ringing noise to the modern squeaks. Direction finding to decide which phone is ringing has become much harder with the new gadgets. (Personal beepers are even worse, but there the beeped person probably prefers not to be identified.)

One of the two persons whose acquaintance I made by this discussion in the faculty club turned out to be a Cornell professor of biology, Tom Eisner, whom I later got to know well, when I was at Cornell. He was engaged in biological work of the highest quality and achieved great distinction.

At the Harvard College Observatory, I was in charge of the radio astronomy program. We had a small but good-quality radio telescope in the countryside nearby. I thought that I could turn this into a major instrument for the new 21-cm observations by building a novel kind of amplifier for it, one of very much higher sensitivity than the standard ones. The maser had recently been invented, which is the radio frequency amplifier that led later to the equivalent amplifier for light, known as the laser. If a maser could be made for 21-cm radiation, then this little instrument would be the equivalent of much larger instruments elsewhere. We knew roughly what we had to do, though the detailed design was still a big problem for us.

I applied for a grant for the design and manufacture of such a device, and in due course, I received it. The money allowed me to set up two scientists familiar with radio techniques, to work on this for a year. I had the good fortune of knowing just the right persons for this: one was John Jelley, a physicist whom I knew from cosmic ray work in England, and the other was Brian Cooper, a microwave engineer from Australia. The expertise to set up the project and to guide it was available in the form of Dr. Nico Bloembergen, a leading Harvard physicist in the subject and a delightful person to work with. It all went as planned, and the Harvard radio telescope became the first instrument to be equipped with such a supersensitive receiver. Of course, I cannot claim any credit for this. I did not do much more than just think how nice it would be to have such a gadget. The credit belongs entirely to Bloembergen, Jelley, and Cooper.

McGeorge Bundy was the dean of arts and sciences at the time (later of fame as chairman of Foundations and as National Security Advisor to Kennedy), and he organized travels to Harvard alumni clubs all over the country. This was always very well orchestrated. Lectures were filmed and edited for wider distribution, and this road show was intended to explain what Harvard was doing in this day. He had selected Purcell and me as the major players in this, and, of course, I was thrilled to do it.

Nevertheless, despite all these attractions, I was not at ease in the Harvard setting. One reason was that I was not really familiar with the whole setting and with its ways, and even though I was clearly well appreciated, I just did not feel at home. Another reason was the amount of wasted time driving to and from work each day, almost 2 h total. Even though the Boston area would have had great attractions, and I would have enjoyed dinners with friends, I could not bring myself to drive into town a second time. Also, not through any fault of Harvard, my life was made miserable by the affliction of recurrent pancreatitis, which laid me low on many days. I traveled a great deal in those days, to conferences, committee meetings, and scientific discussions with friends. One trip was going to take me to Pasadena, and there I came upon a phenomenon like no other.

A Genius with Puzzles

"When you get to Pasadena, you absolutely must meet Feynman," some of my colleagues in Cambridge told me in 1955 as I was preparing this trip. I corresponded with him, and he suggested that we meet for breakfast in a pancake restaurant. We were both into intellectual teasers and puzzles, and our getting acquainted was to exchange all the ones we knew. Of course, he had heard most of mine before, or sufficiently similar ones, so they presented no challenge. Equally, he presented his with the words, "You must know this one," but as with telling old jokes, it is still entertaining among puzzle addicts.

I had invented one myself just shortly before, a rather clever and tricky one, I thought, which none of the Cambridge addicts had been able to do on the spot. He could not have heard that one. As I explained the puzzle, he kept completing my sentences. When the number 26 came up, he said, "27 minus 1, so you can't do it with 27." Before I came to the end, to pose the questions, he had already thought out the elegant solution and, with that, had reinvented the puzzle.

We sat in that restaurant over lunch and, in fact, until evening, enjoying puzzles and subtle problems from the physical sciences. He clearly had thought through a lot of subjects, even much that was quite far removed from his own major activities. Even if something was new to him, it did not take him long to explain it, especially if there was some particular difficulty. His speed was breathtaking!

Many years later, in my house in Ithaca, we got him to discuss his ability with combination locks on safes. (Feynman wrote "I have been here" in a supersecret safe in Los Alamos, just to see whether the owner would own up to security over the infringement!) My 5-year-old daughter brought down her little toy safe with a 3-figure combination lock. "Can you do that one?" she asked.

That would have been quite a problem, if he had really to go through a few hundred combinations. To avoid embarrassment, I tried to divert to another subject. But no. He shook it a little, turned the knob, and listened—just as you see in the movies—and in a moment, it was open! Never mind the 5-figure lock in Los Alamos, in the eyes of my daughter he really was a genuine safecracker.

It is said of some people who are modest that they have a lot to be modest about. Feynman was not modest, but he had more to be immodest about than anyone I have ever come across. I once had occasion to explain to him a subtle point about planetary atmospheres: that if they are very opaque, they will develop the adiabatic temperature gradient even if heated only from above in the uppermost layers, and the bottom will therefore get very hot without infringing any laws of thermodynamics. He said, "Gee! I had never thought of that." Why should he have? It was not near any of his subjects, but it was subtle, and as with puzzles, it was his domain.

Once, some years later in his house in Pasadena, his son remarked to him, "My teacher says you must be an absolute genius." I assume his father had helped with some homework and swindled in some clever trick. That the boy was unaware of the truth of the remark is fine, but did the teacher really not know about whom he was talking? I said to the boy, "Maybe your teacher is right." He replied, "Naaa," and shook his head. To an 11-year-old, a father is a father and not a genius!

But to many of us who knew him, he remains a genius. Not a distant, impenetrable scientific genius as so often portrayed in fiction but that bright Brooklyn boy, immensely proud of his achievements, yet retaining his brimfull measure of good humor, wit, and charm. How marvelous that such an intellect and that personality could meet together in one person!

The Arrow of Time

A topic that had interested me since the days of our discussions of cosmology, and to which I returned in the newly won freedom I had at Harvard, was the nature of time. What was it that gives any one of us the certainty that time flows, and that flow has a clearly defined direction: from the past to the future? The laws of physics do not suggest any such flow, but instead, they are quite time symmetrical. Newton's words had been "we must assume that there exists a mathematical flow of time," but he did not elaborate beyond that.

It must be a universal effect, not one confined to our particular locality, and thus belongs into the discussion of cosmology. We have long known an effect in cosmology that seems to bear on this. We see the stars pouring out their energy of radiation, all of which ends up being sent out into the universe, from where very little of it returns. If all that energy returned, then everything would turn into a uniform heat bath and there would be no statistical distinction between past and future, and so nothing would indicate a flow of time. Yet in that heat bath, the physical laws may still be the same as those we know. For a time asymmetry to be introduced, we must have a universe that can swallow up whatever heat energy is poured out in it. An expanding universe with the expansion at great distances approaching the speed of light represents one clear solution of this problem. The architecture and motion of the universe thus come into the definition of the difference between past and future. I had therefore thought whether this flux of energy into the depths of space could be an effect that also shows itself prominently in all the processes that we observe and that this is then the origin of our clear recognition of the sense of time's flow.

I had never been able to reconcile myself to the notion that the large-scale structure of the universe and the laws of physics it possesses were two independent subjects. Rather I thought that one day we might understand how the laws of physics we know could only exist in a universe like the one we have, or, conversely, that the universe we have would necessarily show us the laws we know. Here we have a case in point that a very basic property of the time-symmetrical physical laws is the difference between past and future, and this is due to the statistical effects resulting from the architecture and motion of the universe.

Newton had said, "we must assume that there exists a mathematical flow of time." He had seen that this was a necessary assumption to give meaning to his physical laws in which he had used time as a coordinate.

In relativity theory, time became interrelated with space and with gravitation—a more complicated concept, but nevertheless, there is still a flow of time. It flows at different speeds in different places, for example, slower by about one part in one million on the Sun; we know this because we can observe by the vibration speed of atoms in the Sun's photosphere. The amount of time elapsed on a fast flight returning to the starting point will not be quite the same as the elapsed time a similar clock would show that had stayed behind. An airplane ride around the Earth with an atomic clock showed this effect. But while Einstein's universe of space and time will be seen quite differently by different observers, moving differently or in different gravitational fields, the concept of a flowing time is still there for all of them. We, and everything around us that we can see, will be moved along the coordinate of time.

The term "basic physical law" refers to the behavior of atoms and all the interactions they display, through force fields we may call nuclear, electromagnetic, and gravitational. But it does not include the statistical consequences that may follow from these same interactions, an important distinction to which we shall return shortly.

Newton's laws of dynamics would apply equally well if we inverted the sense of time in a description of a dynamical process. The motions of the planets, for example, would follow Newton's laws just as well if all motions were reversed. The same is true of all other dynamical processes. When Maxwell developed the laws of electricity and magnetism, similarly there was nothing that defined a direction for time's arrow.

Oh, but is this really so? Don't these laws require the definition of a fundamental constant of nature, the speed of light? Does that not mean that a pulse of light will arrive at a later time than that at which it was emitted—later, according to our concept of the direction of time? Does that not define the arrow of time objectively, independent of our introspective notions? After all, we never see a light pulse arrive before it was emitted.

No, this still does not define time's arrow. Why not? It is because the detailed processes of emission of light and its reception are exactly alike, and what in one sense of time would be called emission would in the opposite sense of time be called reception. If there existed the slightest difference between the processes of emission and reception, then of course time's arrow would be defined by this. But there is not; nature is very careful at every step to avoid showing any preference for one or the other sense.

Quantum theory, the other fundamental set of physical laws, does not resolve the difficulty either. It is said that in quantum theory, the future is defined only in a probabilistic sense. We cannot know what will happen as a result of defined starting conditions of an experiment, but we can only give the probabilities for each of a number of possible outcomes. Does this not distinguish between future and past and thus give a definite sense to the flow of time?

Again, it does not do this. Knowing the configuration of the items in our experiment at one instant allows us only to calculate the probabilities of the various possible configurations at a later time. But it does not give us any different ability for knowing its past. In a system of limited extent, unperturbed by external influences, both the future and the past can be known only by these same probability laws of quantum theory. Thus, if we had some particles in an isolated box, and we knew their masses, positions, and velocities at one instant, we could calculate the probabilities that they would be in certain places with certain velocities at some "future" moment. But if we looked at the same information and believed that time progressed in the opposite sense, then of course we would regard the velocities as the opposite of those in the previous description, and we could now calculate the probabilities that the particles would be in certain positions with certain velocities at some moment we would now regard as "future," but it is a moment "past" in the first mentioned choice of the sense of time. The two calculations would have been different in detail, but they would have been subject to the same probabilistic rules. Time's arrow would not be in evidence. If we had a movie strip taken of these particles, an examination of it would not have allowed us to determine in which sense it had been run through the camera.

But is this not extreme and absurd sophistry? After all, the past has happened, and could have been recorded quite definitively, while the future can have no such record.

But this "definitive" record of the past is always dependent on circumstances outside any limited, defined experiment. It is indeed only a case of "predicting" the past by probabilities, just as it is for "predicting" the future; only for the past we may draw on a much larger array of "information" from sources quite outside our experiment, like the notebook of the experimenter, while for the prediction of the future, we must limit ourselves to the probabilistic information we can get from within our limited experiment.

There is no question that the real world shows the past and the future in a completely different way. It is not that the distinction is one of our own making. Quite objective experiments can be set up that make a clear case that the behavior of the real world is not time symmetrical. Heat flows from a hotter body to a colder one and never in reverse. Any one of the individual atomic collisions involved could have transported energy the other way, and some indeed will have done so. It is only in the probabilities when large numbers are involved that we see a clear asymmetry between past and future. When those numbers are very large, as in the cold and the hot body with their zillions and zillions of particles, the probabilities become what we describe as certainties. The only question is why this is so, when the fundamental laws did not feed any time asymmetry into the system. Where then did it come from? Why only in large systems with many particles like the hot and cold body, or, indeed, our brains?

Suppose that we had some movie films, and the direction in which they had run through the camera was not given. Could we tell? On some, we could with a very high probability of being right, while on others, we could not. It depends what it is that they show. If it was an automobile running along a road, we would say that it was probably running in the forward direction. But of course, it could have been running in reverse, with a very skillful driver. Or a person on a bicycle? Well, it could have been a circus acrobat who had mastered driving his bicycle backwards. But then, we might have the high diver, diving off a high board into the pool. Surely there can be no question. There is no way in which the waves in the pool could have been set up to converge on the submerged person and squeeze him, like a melon seed between fingers, so that he would be flung neatly upward and land gracefully on the high board. If the automobile had been filmed running along a dusty dirt road, we would also know; there is no way in which it could have run backward and vacuumed up all that spread out dust cloud that hung over the road. Or an airplane landing: could an airplane take off flying in reverse?

It is clear that the direction of time's arrow is always defined in complex systems that display the behavior of very many particles: the molecules of the water, the dust grains on the road, and the molecules of the air that support the airplane. Although the collision of individual elementary particles is time symmetrical, in the real world that surrounds them, there are many events that are not. It is not in the fundamental laws that we know today but in their application to "real world" situations that time's arrow becomes evident.

Time is demonstrated only by the statistical properties that systems of some complexity can possess, which are changed only when there is a flow of energy through them. When heat flows from a warmer body to a cooler one, this represents a flow of energy and a change of the temperatures of the bodies. Temperature is a statistically defined property indicating an average of the energy of the thermal motions of the countless numbers of particles that make up the bodies. If we were dealing with bodies composed of a few particles only, such statistical quantities would have little meaning: heat might frequently flow from the cooler to the hotter body. When the numbers are enormous, statistical probabilities turn effectively into certainties. Yes, I could calculate the probability that my hot water tank would get hot just by taking energy from the cold air around it; it would not infringe any basic law of nature. But the odds against this happening are large. We know nothing with greater certainty than that this will not happen.

If we lived in a universe composed of a small number of particles only, there would be no reliable definition of the arrow of time. No time-asymmetrical effects would show up with any great statistical significance. If we had a box that was perfectly insulated from outside influences so that no energy could flow in or out, it would eventually reach such a condition. We might have put a clock into it, but it will eventually run out of the energy it needs to run. But, equally, if we lived in a universe that behaved just like a large version of the heat-insulated box, we have a system in which there is no direction specified for the flow of time. Indeed we could, at least in principle. If we put a hot and a cold body into it, there would be a flow of heat, and the direction of time would therefore still be shown. But also this process would come to an end when everything inside had reached the same temperature. The molecules would still be doing their thermal motions, but we would now find that there is nothing that gives any direction to a flow of time: not the individual processes like collisions between particles or emission and receptions of photons, but now even the statistical properties of vast numbers would be time symmetrical. So it is not just the complexity that gives us the flow of time, but there has to be a flow of energy from a warmer to a colder body. In fact, we live in a universe composed of a very large number of units—particles, photons, or whatever—and containing sources of energy, principally the stars.

But for a flow of energy, we need not only sources but also sinks. The energy sink that is always the final one in any flow is the depth of space. There may be many intermediate steps, but finally, all the energy is tipped into this bottomless pit. All these stars are radiating outwards, and yet the depths of space can accommodate all this radiation without being heated to the temperature of the stars. We suppose that the expansion of the universe is responsible for this sink: whatever radiation energy is received out there is received on bodies that are moving away at a high speed. The energy registered there per unit time is diminished compared with the emitted one. The motion of the objects away from us means that more energy is stored in transit, as the path length to them keeps increasing, and it also means that the radiation received there gives an extra push to the objects there, which will be added to the dynamical energy of the expansion of the universe. What is reflected from them back toward the sources is weakened even further by this same motion. It is the overall expansion of the universe that is responsible for these effects and, therefore, for that black, ever-absorbing sky we see. If the universe were much denser and hence much less transparent, or expanding much more slowly, then it might have filled up with all the radiated energy, and the "heat death" would have arrived. Not even the statistical properties of many particles would show any meaningful difference in the two directions of time. But apparently, the universe is large, transparent enough, and expanding to accept nearly all the energy that is tipped into it. This is then the time asymmetry that we had to find to explain why the time-symmetrical physics still leads to time-asymmetrical consequences.

But can we really implicate the distant universe and its motion to explain effects we can all see around us every second? Of course we can.

Here on Earth we have the Sun, our main source of energy, pouring out radiation energy derived from its nuclear sources inside and sending it out into deep space. A small proportion of it is intercepted by our planet and sustains all kinds of strange occurrences there, ourselves included. Every mechanical motion we can produce, by ourselves or our implements, has used solar heat as the energy source, either directly or through the production of chemical energy we can use. So long as the Sun shines, and so long as the sky can absorb whatever energy is sent out, the Earth will not reach that "heat death" which is the end of the flow of time.

So then here we have a connection between everyday occurrences around us, the little world we live in, and the grand architecture of the cosmos. The most basic property of the physical world, the flow of time, required a setting in a universe that possessed large numbers of particles and had not only sources of energy but also that seemingly infinite sink, so that a flow of energy could continue and show itself as the flow of time on any macroscopic scale. Whether this will be so forever, we do not know.

The topic of the nature of time continues to be an exciting and interesting one, and it may guide us to that next big step in the understanding of the physical world around us. Nature was serving it up to us in an enigmatic form. The darkness of the sky that allows all energy to be deposited, from whatever source it came, and that allows an energy flux to continue to flow from the nuclear energy of the stars and from the gravitational energy possessed by matter into the depths of space gave us all the most primitive and elementary perception of our universe: the flow of time. Just as nature showed us no signs in its basic laws for the quality of inertia (acceleration of a mass, but relative to what?), so it shows us no signs on a microscopic scale for the flow of time. In both cases, we have to look at the large-scale behavior of the actual universe to identify the cause for effects we regard as most basic.

Could there be other universes that work on different principles? We do not yet know the answers to this question, but I tend to think the answer is no. A better understanding will eventually show us a uniqueness for this combination of the local laws and the grand architecture of our universe. If we were to find this, we could have some confidence that we had gained a higher level of understanding of the nature of the Cosmos and of its regularities which we call the laws of physics.

The Solvay Conference

My discussions on the nature of time aroused some interest in physics circles that then resulted in an invitation in 1958 to the Eleventh Solvay Conference, to speak on this topic. The Solvay International Institute of Physics holds meetings at the University of Brussels, on particular topics that are at the forefront of physics at the time, and invites many of the leading physicists of the day. I regarded it as a substantial honor to have been invited. This conference was "On the Nature of the Universe," and my friends Bondi and Hoyle were also invited. Among the other invitees were outstanding persons like the physicist Wolfgang Pauli, the astronomer Walter Baade, the past director of the Manhattan Project Robert Oppenheimer, the Armenian astronomer Viktor Ambartsumian, and many other notables who I first met there but with whom I later had contact over the years. I realized, as I looked around the conference table, that I was probably the youngest person present (Fig. 6.1).

It seems the paper I wrote on this topic of the nature of time for the proceedings of the conference was generally very well received among physicists, and I think it

Fig. 6.1 The Solvay Conference 1958 was attended by Bondi, Gold and Hoyle, the founders of the steady state model, as well as Georges Lemaître and Allan Sandage, promoters of big bang cosmology

became more or less the standard outlook. It was later reprinted in an Einstein memorial publication, and I was invited by the American Physical Society to give the Richtmyer Memorial lecture on the same topic. Thermodynamics came to be regarded as the entry of physics into the asymmetry of time. The Solvay Conference had been a very good and interesting conference, superbly arranged as are all Solvay conferences. For me, the chance to meet these people was a major bonus. Pauli had a great reputation of destroying speakers at conferences by some incisive remark, and I knew that I had to prepare myself carefully for my speech. He sat there listening but did not say a word when I had finished. That, coming from Pauli, I regarded as a compliment.

Chapter 7
Move to Cornell

Setting Up the Center for Radiophysics and Space Research

In the winter of 1959, I received, for the second time, an offer from Cornell University. It was a very attractive offer. I was asked to set up an interdisciplinary center bringing together people from physics, astronomy, and electrical engineering, concerned with radio astronomy and ionospheric research. At the same time, I would be appointed chairman of the Astronomy Department which at that time consisted of just two persons. I was assured that I could make several faculty appointments. There were plans to build the world's largest radio astronomy instrument: a 1,000-ft-diameter dish in a natural bowl in Puerto Rico. It was clear that an instrument of that magnitude would put Cornell in the forefront of radio astronomy, make possible the radar observation of the nearer planets, and allow a new type of observation for the study of the ionosphere. This instrument would clearly be the major piece of equipment of this new center.

I made the decision to accept this, but with a heavy heart. Should I move away from the most prestigious university in the United States to an out-of-the-way, though very good, university in upstate New York? Would I lose all the friends I had made at Harvard and MIT? Would I lose my acquaintance with that rich pool of first-rate physicists and astronomers?

Very few people had ever resigned a Harvard professorship. All the same, the attraction of setting up a major organization to build the world's largest radio instrument was too great an attraction to turn down. In the spring of 1960, we moved back to Ithaca, the small university city, set in quiet and beautiful surroundings.

At Cornell, I immediately went to work to build up my new organization, which I named "The Center for Radiophysics and Space Research." Initially, the new center did not have a building, but I had offices scattered around the campus in Physics, Electrical Engineering, and in the small buildings in which the two initial members of the Astronomy Department were housed. This situation provided me with no doubt very healthful exercise because I ran any number of times each day across the whole

Thomas Gold *edited by* Simon Mitton, *Taking the Back off the Watch*,
Astrophysics and Space Science Library 381, DOI 10.1007/978-3-642-27588-3_7,

Cornell campus to visit the different members of the center. The new appointments I made were, in the first place, people I had come to know as students at Harvard and MIT. Henry Booker, a British scientist whom I had known from wartime days in Britain, was the head of the Department of Electrical Engineering at Cornell. Since the center that I was setting up had a lot to do with electrical engineering, he was appointed the associate director of the center. Another major player at this stage was William Gordon, an electrical engineer who had been responsible for getting the Department of Defense to provide funds for that large radio antenna that was planned to be built at Arecibo in Puerto Rico. I had clearly been given to understand when I accepted the position at Cornell that the construction and operation of this very large radio instrument would be under my control, as director of the center.

The Great Radio Telescope at Arecibo

I had had some slight contact with the initial phases of this instrument when, at a meeting in Boulder, Colorado, 3 years earlier, I had met Bill Gordon, who was then the chairman of a panel set up by the National Science Foundation commissioned to prepare a plan for a large US antenna for radio astronomical purposes. The British instrument at Jodrell Bank, UK, was then the world's largest, and although it had some technical difficulties, it was clearly an enormous success for radio astronomy. I had, since my wartime radar days, a great desire to see the development of radar for planetary astronomy as a way of measuring so much about the planetary system that was otherwise unknown. I was therefore very keen to advise Bill Gordon that the next step after Jodrell Bank would be to build a much larger instrument, so large that one could, indeed, get radar echoes at least from the nearer planets. With the transmitting and receiving equipment available at that time, it would clearly be a dish much too large to be made moveable like the Jodrell Bank instrument in Britain. But it would have to be an instrument that consisted of a large, upward looking dish and some limited capability adjustment to move the radio beam by small angles so as to look at the nearer planets at some time of the year. In addition, such an instrument would have the sensitivity for radio-astronomical observations to a great depth into space, even if only in a limited angle. The number of radio sources that could be investigated for a given price would not be all that different between the two solutions of a smaller moveable instrument and a larger one, of very limited angular coverage. But then, I thought, it would probably be of greater interest to see some sources that were farther away than to see more at nearer distances. So, for both these reasons, I judged that the best instrument to build after Jodrell Bank was a very much larger one but with a fixed, upward looking reflector.

Gordon's interest was mainly in investigations of the Earth's ionosphere, the ionized reflective layer in the outer atmosphere that was, at that time, still a significant factor affecting long-distance radio communications on the Earth. The Department of Defense was interested in such problems as the changes to the ionosphere that nuclear bombs could make. I was keen to see that we got teams

Fig. 7.1 Gold lecturing at the Sydney University Summer School in the mid-1960s

together for radio astronomy and for planetary radar, both subjects for which this instrument would be the leading one in the world. Gordon and I were in agreement that we should plan for roughly one third of the instrument's time to go to each of the three fields: ionospheric physics, planetary radar, and radio astronomy (Fig. 7.1).

The Cornell administration gave Gordon the responsibility for selecting the contractors, for choosing the design, and for administering the construction. This gave me some concerns since I thought that I would still be held responsible for any problems that might arise in design and construction, and I was well aware that quite major problems must be expected in such a large and complex undertaking. When the construction began, Gordon moved to Puerto Rico, and largely cut himself off from Cornell. I was now left without any information about many design decisions, and I also felt that it would be difficult for me in these circumstances to assemble persons with the necessary interests and skills to use the instrument for radio and radar astronomy. The allocation of funds for this purpose was left up in the air, and I was becoming quite concerned that the instrument would be completed but that we would not have the teams to operate it.

I had one stroke of good fortune when I visited a large radio technology site of the Air Force, situated near Cambridge, Massachusetts. Radio technology had advanced considerably since my days with British wartime radar, and I wanted to familiarize myself with these advances. A young man by the name of Gordon Pettengill was delegated to show me around and explain the equipment. He did this superbly well, understood very quickly what I did and did not know, and my few hours there were of the greatest value. I was so impressed with Pettengill and his technical knowledge and skills that I decided he would be the best man to become the Associate Director of the Arecibo telescope, and I felt sure his interests would turn mainly to planetary radar. Bill Gordon was to be the director. When the instrument was completed in 1963, Pettengill indeed accepted the associate director position and moved there to start a vigorous and extremely successful planetary radar program.

For radio astronomy, I had Frank Drake, whom I had known from my days at Harvard as a very good graduate student and a first-rate radio astronomer. I had been able to appoint him to a faculty position in the Astronomy department, and Arecibo was certainly going to be the instrument of his choice.

The instrument was completed in 1963, and though the three subjects, ionosphere, planetary radar, and radio astronomy, each had excellent leaders, there was still a shortage in the United States of skilled personnel in all three fields. The plan had been to invite scientists from other centers to submit proposals for work with the instrument, and government agencies would fund such visits. But far too few proposals were received to occupy the instrument around the clock, as had been intended.

Another problem raised its head once the instrument was working. Its actual efficiency was measured and found to be far below expectation. The amount of power collected from this 1,000-ft-diameter dish turned out to be only about one quarter of the amount of power that was intercepted by it. We knew the power level that should have been in the receiver input when the antenna was pointed at previously well-measured radio sources. And numerous such measurements showed the Arecibo receiver input power to be so far below the expected value that we clearly had to find out why this was the case and what could be done about it. Obviously, an antenna of only one quarter the area would have been very much cheaper to build, and all the extra expense that had gone into this instrument would not be justified until we could fix it (Fig. 7.2).

Fig. 7.2 Thomas Gold and George Gamow photographed by Harry Messel in Sydney on the occasion of their televised debate about rival theories in cosmology

For quite a while, Bill Gordon thought that this was not a major problem, but then perhaps some of the surveying and the positioning of the antenna system that collected the power reflected from the dish to lead it to the receiver input had introduced problems that minor adjustments could fix. I became increasingly dissatisfied with leaving this great instrument in such a debilitated state, and I wanted to make sure that the world's best experts in antenna design were brought in to find the cause and the cure.

The Arecibo instrument was of an unusual design, necessitated by having a fixed reflector. The more usual design had been a moveable reflector, which could then be of parabolic shape, which would then concentrate all the power intercepted by the dish to one point at which the input to the receiver is then located. For a fixed dish that would allow some limited angle of adjustment of the beam, the primary reflector has to be a part of a sphere, and then, the reflected power is not focused onto a point, but onto a line that represents part of a line drawn through the center of that sphere. The power had therefore to be gathered up along this section of a radio line by a device that bears the name of a"line feed." The movement of the beam is then achieved by moving this line feed but such that it remains radial to the spherical reflector. It was somewhere in this complicated system that this large loss of power seemed to occur.

When I had become convinced that minor adjustments would not fix it, I invited Professor Ronald Bracewell from Stanford University to spend a semester at Cornell to go over the calculations and the performance of this line feed. The initial design had been made by an antenna design group in Cambridge, Massachusetts, belonging to the Air Force. I now wanted an independent evaluation. Bracewell was my choice for the first step since I knew him as a person of great understanding and mathematical skills in the fields related to antenna designs. He came and remarkably quickly found the nature of the errors of the existing line feed. There was nothing that could be done by adjustments; it was a matter of designing a completely new line feed. I hunted around for a company that could do the lengthy and quite elaborate work that was needed for the design now that the principles had been clearly laid down by Bracewell. I went to various companies that did work in this field and discussed the matter and finally settled on a first-rate antenna designer, Alan Love, who worked for an electronics company in California. Love quickly produced an outline of a design, having clearly understood Bracewell's deductions. A proposal had to be written up to be submitted to the branch of the Defense Department that had funded Arecibo. That proposal contained Alan Love's outline and the estimated cost of design and fabrication, but it was turned down. The reason given was that we had obviously made a mistake in the first place, and they were not willing to throw good money after bad. I felt this was particularly hurtful since I had not been consulted in the original design, and had, in fact, in the face of some opposition, worked on finding a way of saving the performance of our great instrument. Three more years went by during which the instrument was still limping along with its poor performance, although we had the detailed design that would fix it. It was now 9 years after the completion of the instrument before the proposal was accepted and the design by Love completed and built. For me, the procrastination was a senseless waste of many years, and not a good recommendation for the management of the largest radio instrument in the world. The Love line feed worked faultlessly from the moment it was installed, and the instrument now performed according to initial expectations.

For me, these years had been very hard. I thought many times that it might have been better for me to have declined the Cornell invitation and to have remained at Harvard. But the lure of constructing the world's largest and best—as I thought—radio telescope and to operate it for the observations about which I had dreamed for a long time was a prize I could not reject. But as it turned out, the Cornell administration was quite unwilling to come through on the commitment they had made to me. The construction was not put in my hands, and the design decisions that were made were not even shown to me before they were implemented. Nevertheless, the serious fault of the first line feed was then, eventually, attributed to me, and none of those responsible for it ever explained that I had been prevented from even examining the design. Would I have done better? I won't claim that, although I had some education in antenna design in wartime radar. But I would certainly have involved some of the experts I knew outside the Air Force Cambridge group, both before the final manufacture and also after the initial unsatisfactory performance had become clear.

In late 1959, I had come to Cornell with a high reputation. Now, 15 years later, I was widely criticized on several fronts: I had erred in the design of the Arecibo telescope, and I had failed to employ it for the prime purpose for which it had been funded, the investigation of the ionosphere. The administration of the instrument was in disarray. The inadequate use of the telescope for ionospheric researches was constantly attributed to my preference for radio astronomy and planetary radar. In reality, my preferences had nothing to do with the fraction of instrument time allotted to each field. It proved to be very difficult to keep the instrument occupied around the clock with projects run by persons competent in any of the three fields. Pettengill and persons he had selected, chief among them Rolf Dyce from the Stanford Laboratories and later Don Campbell, were doing a superb job in planetary radar. But the positions of the planets and the limited angular coverage of the telescope meant that this subject could not be a major user of telescope time. Radio astronomy could provide a huge number of projects for which the telescope was the world's best, but the number of radio astronomers in the United States was still very small, and not many came forward with proposals to use available observing time at Arecibo. The few radio astronomy centers that could have provided us with investigators mostly had their own instruments and their staff members were fully occupied with those. At Cornell, I had appointed Frank Drake in this field, but I could not make any other radio astronomy appointment, and one person's projects could not occupy much of the instrument time.

For ionospheric work, the situation was even worse. Although funding was available for projects in this field, more so than for the other two, only very few applications were received. There were just not many ionospheric investigators, and the existence of the Arecibo telescope could not suddenly increase that number. Only one application for ionospheric observing time and funding was ever turned down in the 10 years in which I was responsible for the overall control and that was of a proposal judged to be inadequate by Gordon and other ionospheric experts. Yet I suffered the constant criticism for more observing time being used for the other two subjects and too little for the ionospheric work which had been the main purpose of the construction of the instrument.

There was one country that had a surplus of excellent scientists in the field, and that was Australia. Radio astronomy owed a lot to Australia. Many leading scientists in the field were Australians: Bernie Mills, Bill (Wilbur) Christianson, Ronald Bracewell, Joe Pawsey, and several others. They had built instruments of novel designs and attracted the best graduate students. When it became clear to me that the real difficulty for Arecibo was finding scientists who could utilize the instrument to the limits of its capability, I looked to Australia.

The physics department at the University of Sydney, chaired by Professor Harry Messel, invited me to be part of the Sydney Summer School lecture series. I traveled to Australia many times at their invitation. These were lectures on many topics, given by invited lecturers mostly from abroad and high school students from many countries, selected for their outstanding achievements. It was an enterprise set up by Harry Messel, for which he had obtained the support of foreign governments and of leading persons in Australian industry. The program proved to be an enormous success, and I enjoyed being a part of it. During these visits, I came to know the radio astronomers and their students, many at the University of Sydney. When Harry Messel heard of my difficulties in staffing Arecibo, he made a great suggestion: we could form an association between our two organizations, within which we could freely interchange personnel. A large amount of administrative work was required to allow Australian scientists access to Arecibo and to pay for their travel expenses. But all this was done, and the Cornell-Sydney University Astronomy Center was established in 1964. It was a great success. For several years, many excellent scientists, who enjoyed working the great instrument at Arecibo, performed a large portion of the types of observations for which Arecibo was the world's outstanding instrument (Fig. 7.3).

Fig. 7.3 Harry Messel presents a gift of an aboriginal bark painting to Tommy Gold on occasion of his lecturing for the Sydney University Summer School

There were many observations that benefited from comparisons between the Northern Hemisphere that Arecibo could see and corresponding observations in the Southern Hemisphere accessible to our partners with their large instruments there. For radio astronomy and planetary radar work, Arecibo was now in full swing. But for the ionospheric work, we still had too few applicants. This caused continued criticism, as if by better time allocation, I could have produced applicants that in fact did not exist.

The Cornell-Sydney University Astronomy Center worked well for about 4 years, but then the National Science Foundation (US) decided that we had no right to make our (US funded) instrument available to Australians, and this partnership had to be abandoned. The work at Arecibo had benefited from the work of several Australian scientists, and we retained a number of them on the Cornell staff.

My frequent trips to Sydney as a lecturer in the Summer School series proved to be very enjoyable. The lectures on diverse subjects were sent out over national television channels and well received by the public. I was even told, as the ultimate compliment, that the listener survey had shown a higher attendance for these lectures than for the television series, "Gunsmoke."

The most memorable lecture in this series was a one and a half hour debate between me and George Gamow, a highly respected physicist. Gamow had been the driving force behind the big bang theories of the universe, according to which it came into existence in a sudden event and then blew itself up to the large size we now see. I had been the proponent of the steady state universe, and the debate between those two viewpoints brought out all the advantages and disadvantages of each. This apparently was a great occasion for the Australian television system, and I recall being asked by strangers on the street about individual items of cosmological theory.

But my visits to Australia were not confined to giving lectures and selecting students to work temporarily with us. Traveling to Sydney allowed me to take stopovers on various occasions in Hawaii, Tahiti, Hong Kong, and Singapore. Those were all instructive and also very enjoyable. In Australia, Harry Messel took me for long walks through the beautiful Blue Mountains, through rain forests, and into the deserts of the "outback." On many weekends, he invited me to his beautiful summer home on the Hawkesbury River, and there I came to see an aspect of the "Aussie" culture—their toughness, their humor, and their delight in sports and fun.

The Hawkesbury River, a little north of Sydney, is the Mecca of water-skiing. Harry had a powerful boat right by his house, and we enjoyed many hours of water-skiing and learning all kinds of tricks. The best, however, was the display of those who could really do it, who usually used the early morning hours when the river water was absolutely calm. There you could see people barefoot skiing on one foot only (barefoot skiing on two feet was a thing of the past). They would use the other foot to hold the tow line. But then there was the even greater surprise: some of them could freely jump around to go backward, still not losing the foothold on the tow rope. As an exhibition over breakfast, it was superb. But perhaps the most entertaining were the antics of "the butcher."

He was a real butcher in Sydney who had successfully climbed the butchers' ladder to become the chief supplier to the best hotels in Sydney (and to Harry Messel). He also had a country house on the river, a very powerful boat, and a skillful driver, trained to grasp all instructions from the butcher, his master on a water ski. The butcher was quite a character on the Hawkesbury. He was a man of powerful physique, a superb and wild water-skier, and also a good prankster.

The funniest prank we witnessed went as follows: his boat was overtaking that of a young lady skiing quite respectably (on one ski as was the rule), and the butcher gave a signal to his driver. This resulted in his boat being driven alongside the young lady's boat, allowing the butcher to ski quite close to her. Then, in one brief moment, he stepped off his ski and transferred himself onto her ski, put his foot behind her back foot and his other foot in front of her front foot, one hand on her towline and the other holding her firmly around the waist. Of course, he stood on her ski without a binding to give him any grip. The girl shrieked with panic and excitement, as he was now firmly in control of her ski. Then, he did the wildest turns, shoulders almost touching the water, turns which I am sure she would never have been able to do by herself. It was the best quick instruction I could imagine.

His driver, who was quite blasé about the antics of his master, had meanwhile picked up his abandoned ski and positioned it ready for boarding and continuation of the day's exercises. I later found out that this episode had been a way of introducing himself to a neighbor with whom he had previously only exchanged hand-wave greetings.

My antics in water-skiing on the river went as far as skiing on a plain fence board. I asked Harry one day to find me some old board, about 4 or 5 ft long, and I would see whether I could ski on it without binding, or any attachment. And, of course, it would be without the tail fin that is a major stabilizer on the usual single water ski. Harry found me a splintery, rotten old board. I skied on it, crossed the steep wake of his boat and did all the usual things, except of course that I could not jump over the wake since without a binding, I could not arrange that the board would land on the other side of the wake at the right time and place for me to connect up with it again. Harry was very pleased with this performance and kept the fence board in a special place in his shed marked, "visitor's ski." One day, when the butcher came to visit us, Harry showed him this ski and mentioned why it was named the visitor's ski. The butcher looked at it and said, "Oh that's nothing, I've skied on a toilet seat."

Another encounter with the butcher turned out to be somewhat frightening. He had bought himself a tow plane, towed behind his boat, but with a very long cord (I think about 300 ft). It was a light canvas plane, single seat, controlled by a joystick and rudder bar, much like a small airplane. It had water-ski-shaped floats as the landing gear. When he turned up with this contraption, he asked me whether I knew how to fly a plane. I said yes, indeed I had learnt to fly little planes. Then he said, "You must try this one, it is great fun." I agreed to this invitation and installed myself in the tow plane. Before we started he said that there were two things that were very important (1) never to go down so steeply that the plane overshoots the tow rope because if the tow rope goes slack, it will later tighten up suddenly and probably break the attachment and (2) a few miles down the river, there were electric high tension lines and I absolutely must land on the river well before we got to them. The butcher would wave a signal from the boat when it was time to come down.

So the boat started to pull me and almost immediately my plane reached flying speed. I pulled back the stick and the plane absolutely shot up, and then settled to a height of 150 ft. The boat and the people in it looked very small and far away and I could fly the plane quite a ways over the land on each side of the river.

Eventually, the time came when the butcher made a signal for me to come down. I knew that I had to put the nose down very gently, for fear of overshooting the towline. I managed to get myself down to the river safely, but not yet touching the water. In that situation, if one were flying a small plane, one would throttle back, and raise the nose slightly. But in this case, the throttle was not in my plane, it was in the boat. Without the speed diminishing, I was immediately sent back to the height I had flown before. Now the butcher's signals became determined that I must get down very quickly. When I got back to gliding just above the water, I knew that I must not raise the nose again, and I considered that it was better to slam into the water than to fry in the high voltage lines. So I kept the nose down and had quite an impact with the water, but the plane held up. The butcher rolled with laughter when he came back to pick me up and realized how terrified I had been.

The Rotation of the Planet Mercury: A Cautionary Tale

One of the earliest successes of radar astronomy with the Arecibo telescope was the investigation of the rotation speed of the innermost planet, Mercury. Pettengill had perfected the techniques of measuring planetary rotation speeds from the frequency shifts that the echo contained since one side of the planet would be moving toward us and the other side away from us (after allowance is made for the overall motion of the planet relative to that of the earth).

Mercury and Venus were the easiest targets, so they were investigated first by this technique. In the case of Mercury, a remarkable situation developed.

Mercury takes 88 days to complete an orbit around the Sun. It is not a convenient object for optical telescope observations because it is generally too close to the Sun, and the scattered sunlight in our atmosphere is then disturbing. Nevertheless, Mercury had been observed as best one could since the 1880s, and on the basis of these observations, it had been concluded that its rotation was synchronous with its orbit, meaning that it turns once on its axis as it goes around once on its orbit, having one side always facing the Sun and the other side always facing away from the Sun. This state of synchronous rotation is well known as being the motion of the Moon around the Earth. The reason for this is that any asymmetry of the mass distribution or level of the surface of the Moon would capture the Moon into this motion in which the Moon would then not experience any tides. In any other motion, the Moon would have a solid-body tidal wave that would bulge it out a little both in the direction toward the Earth and in the diametrically opposite position away from the Earth. There would be energy dissipated by the internal movement of the body, which is not completely rigid, and this energy would come from the rotational energy that would then be driven to the synchronous motion. It seemed a very obvious guess that Mercury would have fallen into synchronous rotation with its orbit.

When Pettengill made the radar measurement of the speed of rotation of Mercury, he did not find the answer to be 88 days, but instead he found it to be 59 days only, with an accuracy of ±5 days. When this result was announced, the optical observers who had concerned themselves with this problem declared that our radar results must be wrong. They had clear proof of many observations done since the 1880s that showed features on the surface of Mercury that repeated frequently just exactly in the positions in which they would have been for the case of synchronous rotation.

The planet is too close to the Sun to get any good photography, but the various observers had always made a sketch of major features (which we now know to have been the largest craters), and the repeated sketches were frequently in very precise accord with the 88-day rotation rate. Some were not, but as the seeing conditions were not always good, they were considered irrelevant. If some observations fitted the precise scheme that had been predicted anyway, then surely they should be taken more seriously than others that might be less precise. This was the opinion for many years, and the Pettengill observations clearly upset these conclusions.

It so happened that there was a meeting of planetary astronomers from all over the world in the spring of 1965. Audoin Dollfus, a prominent French astronomer, presented the case for the 88-day revolution period that had been certified by these observations for many years. He showed the very detailed sketches, which obviously seemed to support the theory that in every rotation in its orbit, Mercury showed us the same features. On the other hand, I came to this with radar evidence that was clearly in conflict. There is a general tendency to regard new evidence with more suspicion, and so I had to plead for this evidence to be taken seriously. Even though the radar results were much more secure than the sketches of the appearance of Mercury, most observers at this conference seemed to conclude that such remarkably similar sketches could not be attributed to coincidence and therefore that the rotation must indeed have been 88 days as they had expected all along.

When I returned from this conference, it became clear to me that the new radar data that we had were indeed the most accurate that anybody could have obtained for the rotation speed of Mercury. The discrepancy between the optical and the radar observations thus had to lie in the interpretations of one of the two.

The only possibility, or so it seemed to me, was that it had not been correct to state that Mercury would be driven by tidal solid-body friction into a synchronous rotation with the orbit around the Sun. But how could this be wrong? A graduate student of mine, Stan Peale, and I then thought this out in more detail, and we concluded that the tidal friction of the rather elliptic and therefore eccentric orbit of Mercury would be much greater in the sector of the orbit that was closest to the Sun than in all the rest of the orbit. We calculated this in detail on the basis that the spinning speed of Mercury would be stabilized at that speed at which the acceleration in the parts of the orbit nearest the Sun would just balance out the deceleration in the much longer part of the orbit in which the tidal forces were much weaker. This calculation led to 59 days approximately for the rotation period that would be derived by the time-average of these forces.

This theoretically derived period was very close to the period that Pettengill had measured. It seemed to Stan Peale and me that we had hit on the right solution and that the planet was doing just what we should have expected of it. It was spinning at the speed at which acceleration and deceleration around the orbit just balanced out exactly. That indeed was very close to the speed that Pettengill had given us.

The paper we wrote on this subject, which was published in Nature, explains that a planet that has no particular features that prefer any direction of orientation would fall into the pattern of rotation that seemed to be the one we had observed.

What we had said was quite correct. However, we had overlooked an important fact. Shortly after the publication of our paper, Giuseppi Colombo, a celestial mechanics expert at MIT, demonstrated that the figure for the rotation period that Pettengill had obtained was very close to the rotation period that would be enforced by a body that had a slight but permanent asymmetry in mass distribution in its longitude. While we had thought that we had calculated the rotation speed that a body would obtain if it was of quite symmetrical distribution of its masses, Colombo had shown that a body would be captured by the tidal forces just in the rotation speed that we had observed. If the planet rotated at the synchronous speed with its orbit, then every picture of its surface features that is one or an integral number of rotation periods away from an earlier one, and corrected for the different aspect angle in which it is seen from the Earth, should be similar. Many pictures indeed showed just this result. Some did not, but then, this was attributed to seeing conditions inadequate for seeing the details. Surely one thought that if some figures fitted very precisely, we could ignore the ones that did not. How could we suspect that a large fraction would be exactly right for the presupposed rotation period, if that was not the correct story?

What Colombo had proposed was a rotation speed at which the planet would do one and a half turns in the time of its orbit. That comes to about 59 days, very close to Pettengill's figure. The figure that Peale and I had calculated came very close to the same figure, but we had failed to recognize how close this was to the precise synchronism with the orbit, but not with one but with one and a half turns of rotation in each orbit. This would be one of the conditions that are called a spin–orbit resonance. We had particularly concerned ourselves with such things as the orbital resonances of the moons of the major planets, and my student, Peter Goldreich, had greatly clarified the subject. We should have been the first ones to recognize that 59 is very close to two thirds of 88.

What the physical explanation has to be is that the tidal friction, as Peale and I had calculated, brought the spin very close to this value, and then any asymmetry in the mass distribution would cause the spin to lock into this resonance at which any mass surplus meridian of the planet would either face the Sun or face away from the Sun at the nearest approach in each orbit. The tidal force that the Sun exerts is a maximum for each of the two cases. What do we then make of the optical observations that seemed to indicate the 88-day spin? What appears to have happened is that many observations were made when Mercury was furthest from the Sun just because it was then most clearly visible.

Let us number the orbits arbitrarily as a succession of even and odd numbers. In all even numbered orbits, the surface markings would have been the same and similar on all odd numbered orbits. But going from an even to an odd numbered orbit, the markings would have been quite different. It seems that once one set of markings had been adopted as clearly repeating, then any failure to repeat these was regarded as due to inadequate seeing conditions. By this argument, half of the observations must have been rejected. Nobody had thought that this could influence the final decision. This story serves to demonstrate how carefully one has to think through a problem before rejecting any observational data.

Chapter 8
The Pulsar Era

Twinkle, twinkle little star,
how I wonder what you are.
Trad.

It twinkled more than any other star known, and it was smaller than most astronomers had ever suspected, and the world of astronomy wondered what they were. The twinkling was in 1.4-s intervals, the size of the star between 10 and 30 km, as compared with a "normal" star like the sun, whose diameter is about 1,400,000 km. Astronomers worldwide were in a frenzy.

Toward the end of February 1968, a dramatic news item appeared on my desk. The journal Nature published an article on an astronomical discovery that was surely in the first rank of importance. The Cambridge radio-astronomy group had detected rapidly pulsing radio signals from a source in the sky. The pulses seemed to repeat with a precise period of 1.337 s, and the pulses were very short, only a fraction of a second in length. The signal had been observed for several months, and it was clear that it was indeed a celestial signal and not some man-made interference because its position had moved with the sidereal motion of the stars and not with the time of day, which, of course, relates to the position of the Sun. A young graduate student, Miss Jocelyn Bell, had made the discovery as a by-product of other observations of quasars she had been instructed to make by her research supervisor, Professor Antony Hewish. She had the perseverance to follow these strange, pulsed signals that recurred for a long enough time to show that they appeared fixed in the sky and not on the Earth. The individual pulses were short but quite variable in shape and strength and in the precise timing from one to the next. In a narrow frequency band, the pulse length was as short as 16 ms. The long-term mean pulse repetition frequency was remarkably precise.

The range of frequencies over which the pulses could be seen was very large, and it was therefore possible to demonstrate immediately that in each pulse, the low frequencies were quite significantly delayed relative to the high frequencies. This is a well-known effect caused by the presence of ionized gas in space, which makes the low-frequency propagation speed significantly slower than that of the high

Thomas Gold *edited by* Simon Mitton, *Taking the Back off the Watch*,
Astrophysics and Space Science Library 381, DOI 10.1007/978-3-642-27588-3_8,

frequencies. (Very high frequencies like light are almost unaffected, and travel at the vacuum speed of light.) That time difference could be interpreted, with the aid of an approximate knowledge of the electron density in the intervening space, as an indication of the path length from the source to us. It was so interpreted by the Cambridge astronomers, and because there was a fairly good value for this electron density from other observations, the Cambridge group estimated the distance fairly securely at about 200 light-years.

Even if one regarded this determination of distance as only approximate, it was immediately clear that the discovery was showing us something completely new. How could an object at such a distance transmit enough power in pulses as short as a few milliseconds to be observable from Earth? Obviously, the source region had to be small enough so that the differences of flight times of light coming to us from different parts of the region would not exceed a few milliseconds. I calculated what temperature an electron gas would have to have to radiate with that intensity from such a limited volume. The answer was clearly an absurd quantity, far beyond the temperature range that any electron gas could possess anywhere. How else can such intensities be produced? Radio signals in astronomy were generally thought to be produced by a random superposition of the emission from individual electrons, and then the temperature of the electron gas is one way of describing this situation. But in this case, that explanation was clearly wrong. There was no way in which a temperature of 1,025 K could exist in an electron gas since such a gas would make itself known in many other ways, not just by the emission of radio pulses.

But what was wrong with this explanation? If individual electrons were moving independently of each other, the energy emitted would be just proportional to the number of electrons involved. If, however, a large number of electrons moved as units, in synchronism with each other, then the story is quite different. The radiation that results then is proportional to the square of the number of charges in such a unit. Obviously, if in this case independent radiators could not do the job, we had to think of a carefully orchestrated synchronized motion of very many charges.

Radio transmitter tubes, of course, do just that, as I knew full well from my stint of designing some. There, great ingenuity goes into making very many electrons move together. My conclusion was that we had here something that was more similar to such a carefully designed transmitter tube than to a hot electron gas. The number of electronic charges that had to be emitting in a coherent phase had to be at least of order 10^{15} to bring the required velocities down to some acceptable value.

Now, with modern observations on some pulsars, this figure has become more impressive still: at least 10^{18} charges have to radiate in unison. The amount of energy radiated out in high-frequency radio waves, from each square centimeter of the emitting surface at the pulsar, has to be as high as the entire energy output of all the electric power stations on the Earth!

Our Arecibo radio observatory in Puerto Rico had the world's largest and most sensitive antenna. My colleagues and I immediately realized that we should seize our opportunity to use it for pulsar observations. Clearly, it was vastly better suited to observe these signals than was the Cambridge instrument, although, since it could see only a small area of the sky at one time, it would be very time consuming to use it as a search instrument but highly effective as a follow-up instrument for discoveries made elsewhere. The Cambridge discovery paper had given the position in the sky to good accuracy, and so the Arecibo antenna was immediately used to study the pulsating radio source in great detail. It was a straightforward matter to obtain much more detailed information than had been possible in Cambridge.

Soon the news reached us that the Cambridge astronomers had in fact detected three other similar sources, but with different pulse repetition frequencies. Should not Arecibo be engaged immediately to study these? But unfortunately Martin Ryle's secretive Cambridge group was not willing to divulge the positions at this early stage, and Arecibo was not suitable for repeating their search.

It so happened that I was visiting Cambridge at that time, and I brought with me some of the excellent data we had obtained on the first of these sources at Arecibo to show to the Cambridge group. I went to the tea room at the Cavendish Laboratory, a place very familiar to me, and expected to meet the radio astronomers there. They came, and we sat around the table discussing the data. Martin Ryle, joined us, and I showed him our data. "Very good," he said, "it confirms what we have." "Martin," I said, "why don't you give us the positions of the other three sources you have, and we will give you equally good data for them within a few days, if they are in the limited area of sky that Arecibo can see. I promise we will not publish anything about them except as joint authors, and with your agreement." Ryle sat still for a while, contemplating my proposition. It was clearly a tense moment, and his colleagues did not say a word. Then Ryle said, with emphasis in his voice, "No, we will release the positions when we are ready." "But this will delay the time when good Arecibo data will be available," I said. "We have to take that risk" was his reply (Figs. 8.1 and 8.2).

By the evening of that day, I had a phone call to my room in Trinity College from a radio astronomer who evidently had the data and who had overheard or been informed of my request. He gave me the three positions, and I immediately wired them to Arecibo. Evidently this person (I have a suspicion who he was, but no certainty) thought that Ryle's refusal was unreasonable, or at least detrimental to the science, so he broke the embargo Ryle had imposed. At Arecibo, the sources were quickly found and investigated.

Obviously, many astronomers the world over would try to find an explanation for this phenomenon. And so indeed, they did. I was fortunate, for I entered this race with a big advantage. I had thought about just this problem in another context 17 years earlier, when it had been a problem of accounting for the "ordinary" radio sources then seen by the Cambridge radio astronomy group. At the London conference I mentioned earlier, I had argued that small condensed stars with enormously strong magnetic fields would be the only objects of stellar masses that I could understand to give out such intense radio radiation. But, I had added, such objects should show time fluctuations on the short timescale given by the flight time of light across the emitting regions, and no such fluctuations existed for the sources under debate then. So now I felt sure immediately that I knew the answer: these were the objects I had discussed then. The sources must be neutron stars, these enormously dense configurations of matter that the theoreticians had predicted, that

Fig. 8.1 Tommy, in his Doctor's scarlet dress gown, and his second wife Carrie on the occasion of a Feast at Trinity College Cambridge 1971

Fig. 8.2 Tommy and his daughter Lauren in Cambridge, September 1986. Lauren is wearing the uniform of the Perse School for Girls

might pack the mass of the sun into a sphere of perhaps 20-km diameter. They would most probably have intense magnetic fields extending out from their dense surfaces, and with the strong gravitational fields, the atmospheres would be of small height, and the magnetic fields would therefore extend out into low-density space. These were the conditions for optimum efficiency of radio emission, as I had argued. It seemed clear to me now that the objects I had thought about then had now been discovered.

But why did the long-term repetition frequencies of the pulses keep such precise time when the individual pulses were quite inexact? And how did a neutron star keep large numbers of electrons in step to radiate coherently? These were the questions I felt I had to answer.

For an object emitting pulses, as for example by radial pulsations, I thought that each pulse would arise at a time defined by a certain delay time after the previous one. If there was some scatter in that delay time, the long-term accuracy of timekeeping would be given by the statistical additions of the individual pulse-to-pulse variations (a one-dimensional random walk problem, as the mathematicians would call it). But the observed long-term accuracy of these clocks was far superior to that. That was immediately obvious, even on the first records. There was an underlying precise timekeeping that took no notice of the individual pulse "errors." The objects were not pulsing, I concluded, and the name "pulsar" was a misnomer. What appeared here as a pulse must be the sweep of a signal from a rotating beacon. The neutron star would rotate with a very precisely defined frequency, being a very massive object subject to no significant external forces. But the pulses we see would be due to plasma physical phenomena in its surroundings, and variations, both in the strength of each pulse and in the precise positioning of the emitting cloud relative to the star, could well be expected.

So the general picture emerged of a rotating neutron star that possessed some asymmetry around its rotational axis so that some regions of longitude became strong radiators. Perhaps the field was sticking out further there, or it had a greater strength. Perhaps there was a greater supply of charged particles at some longitude than at others. After all, the Sun does a similar thing in its own small way with its regions of sunspots. In the violent collapse which was thought to be the process that leads to neutron stars, the "supernova event," initially unsymmetrical fields, as they exist in stars, would probably be further disturbed, and they would certainly be greatly compressed by an average amount which one could calculate approximately. With such a picture, one could see that the precise timing in the long run would depend on the spin of the neutron star. I gave a comparison of this with an old-fashioned lighthouse that might have a vertical shaft rotating at a well-determined speed but then have a lamp hanging on it from a hook. An observer would see a pulse as the beam from the lamp sweeps over him, but the accuracy of timing he would see might have fluctuations, as the lamp might wobble on its hook. The long-term timing accuracy would, however, be inevitably that of the rotation of the shaft.

This obviously fitted the data very well and persuaded me at the time that I had the correct explanation. But what about the problem of the radiation intensity?

I thought, and my experience in radio tube design had taught me, that a high degree of coherence of the radiating charges could only be achieved if they were all brought to an accurately common velocity. If not, then they would of course all radiate in different phases, and we could not get the large coherence effect. If charges were accelerated, as they are in solar flares (the explosions that happen in the magnetic fields of sunspots), they are accelerated to a great variety of velocities. If, however, they were charges entrapped in a fast-rotating magnetic field, they would all move together, as the rotation enforced. We had long understood that a strong magnetic field of a rotating object would make charges "corotate" out to some distance from the object. What would this distance be in the present case?

Relativity theory is very clear about the impossibility of accelerating any object to a speed greater than that of light. The apparent mass of the object would become infinite, and for a charged object, the radiation it would emit would have a reaction that would hold it back as it approaches the light velocity. Corotation around this neutron star of my imagination would clearly have to cease at that distance at which it would imply motion at the speed of light. This, of course, defines the surface of a cylinder whose radius is given by the rotation speed and the speed of light in free space. The travel time at the speed of light around this cylinder would be the time it took for one rotation of the star. If the star rotated once every second, then this cylinder would have a radius of 48,000 km, and correspondingly less for faster pulsars. I referred to this as the "velocity of light cylinder" or, briefly, the "light cylinder." What I imagined then was that within the light cylinder around a pulsar, charges were whirling around almost as if they were fixed relative to the neutron star. At any one distance, they would therefore move at an accurately defined speed and all together. This, I thought, would fulfill the first requirement for the production of coherent radiation.

Of course, other astronomical objects have a light cylinder also; only its radius is so large that it has no particular importance. No magnetic fields of significant strength stretch out that far from the object to enforce corotation. The Earth's magnetic field, for example, only dominates the external gas to a distance of about 10 Earth's radii. But for the slowly rotating Earth, the light cylinder would be at a distance of 600,000 radii, or 4,000 million km, which is about the distance to the planet Neptune. Obviously, for the Earth, the light cylinder is not a concept of any significance. For the neutron stars, the situation is quite different. With a strong field at the neutron star, corotation of the clouds of charged particles could certainly be enforced out to the light cylinder of a 1-s pulsar. (For a millisecond rotation period, such as that of some recently discovered pulsars, the light cylinder would only be 48 km in radius.) These then were the circumstances that seemed to me to fit what was required for the pulsar explanation.

Neutron stars had been a figment of the imagination of the theoreticians. In 1932, Lev Landau, the great theoretical physicist in Moscow, had made some calculations about matter at nuclear densities making up the cores of massive stars. The astronomers Walter Baade and Fritz Zwicky had speculated that a supernova explosion, the most gigantic fireworks display that astronomers can see in which, for a brief time, a single star may outshine the entire galaxy of stars in which it lives,

Fig. 8.3 Tommy with Frank Drake at the Rome meeting on pulsars December 1969

may signify the collapse of a star down to nuclear density, the neutron star. That configuration of matter is one in which repulsion between neighboring nuclear particles is the only force preventing further collapse. It would be the last stage in the evolution of stars, where all other sources of internal energy that might hold matter apart against its self-gravitation had been exhausted. In that case, a large proportion of the nuclear particles would be turned into neutrons, hence the name neutron star. The density of the object could be even higher than the densities of nuclear matter in heavy atoms and reach incredible values of the order of 10^{15} g per cubic centimeter. A sugar cube piece of it would weigh one billion tons, which is about as much as a small mountain (Fig. 8.3).

Back in 1939, J. Robert Oppenheimer and George Volkoff, following Landau, had made more detailed calculations of the masses and sizes such objects might have. They concluded that masses of one or two solar masses might be involved, and the objects might have sizes between 10 and 100 km. Most astronomers thought that even if they existed, there would be no way in which they could ever be discovered. Objects as small as that could not radiate enough energy from their surfaces to be visible over interstellar distances, even if they were at the highest temperatures one could reasonably envision. The whole concept was not taken very seriously in the astronomical world. I suppose that there were few astronomers who even knew about these theoretical speculations. But no consideration had been given to the two aspects that in the end made them observable, namely, the strong magnetic fields which compression would give them, and the large amount of rotational energy which they would acquire in collapsing from perfectly ordinarily rotating stars down to the small size. I know that one always has to explain that by the example of the skater who starts with a moderate rotation speed with the arms outstretched, and then, by pulling in the arms, spins up to a high speed. For the neutron star collapse, this effect would be so large that the energy content in rotation afterward, all derived from its self-gravity, could well be as large as the entire energy content that the star possessed earlier in the form of all the nuclear energy store with which it started its life. Now the suggestion was that the ability to radiate in the radio spectrum was due to the strong magnetic fields, and the energy source for the radiation would come from the rotational energy.

This then was my model for the process. I had made some calculations as to the strength of the field that could be expected. The astonishing answer was that if some perfectly ordinary star were to collapse, and compress thereby its normal magnetic fields, the magnetic field strength might go up to 10^{12} or 10^{13} G, an unimaginably strong magnetic field. The strongest fields, by comparison, that have been created in the laboratory, even if only for short periods of time, are "only" 10^6 G. What I was talking about now was more than a million times stronger than the strongest man-made magnetic fields! I wasn't sure in the first place whether such a thing was at all possible. Would there be some limit, which the theoreticians dealing with quantum electrodynamics would know, that would prevent such strong fields from ever being set up? I thought, before publishing such a wild notion, I had better inquire whether some other phenomenon would set in that would prevent the formation of such fields. I phoned Murray Gell-Mann at Caltech and presented him with my dilemma. Would there be spontaneous production of particles before such field strengths are reached? Murray thought about it a little and then said he would phone me back with the answer; it was not entirely obvious. When he phoned back, he said I need not worry. The only limits he could see were in the region of 10^{24} gauss, much higher values than I needed for my interpretation.

Many other quantities were in the range of the unbelievable. A density of matter of 10^{15} g/cm^3 is so extraordinary to our conventional thought that it needed an example to bring it home. At a conference in Pisa in 1968, I had the chance of discussing this. My recollection is that I first mentioned the historical report (probably in detail incorrect) that Galileo had dropped two objects of different weights from the Leaning Tower of Pisa (which we could see out of the window) and that he had observed that they hit the grass below at almost exactly the same time; that was obviously in conflict with Aristotle's prediction that bodies fall at a speed proportional to their mass. I explained that if one were to repeat that experiment but take a sugar cube of neutron star material and drop it from the tower, together with a sugar cube of ordinary materials, the two would still hit the grass below at about the same time. But then the difference would become apparent. The ordinary sugar cube would lie on the grass, but the neutron star cube would penetrate into the ground. In fact it would penetrate through the Earth almost as if there was nothing there, and in oscillating in the Earth's gravitational field, it would reappear and come out to much the same height as that from which it had been dropped. The Earth is almost like a good vacuum for such an object. I also mentioned incidentally that there were problems for performing this experiment: firstly, a small object at such a density is not stable, even if anyone knew how to make it; secondly, the leaning tower in Pisa would not be nearly strong enough to support this sugar cube of neutron star material, as it would weigh one billion tons.

I submitted a letter to Nature with my proposed explanation of the nature of pulsars. It was received there on May 20, 1968, explaining these points and predicting from them that pulsars might be found in the location of supernova explosions, following the prediction of Baade and Zwicky. I also noted that neutron stars could rotate considerably faster than the pulsars so far discovered, and, therefore, one should search for shorter periods, perhaps even shorter than 10 ms. Moreover, if rotation was the energy source, one should be able to observe a slight slowdown in this long-term clock. If anything else had been the energy source, then taking energy away from the system would increase the density, and that would make any other periodicity shorter, not longer. So this was the fantastic picture I had conjured up: a mass as large as the mass of the Sun, but only about 20 km or so in diameter, revolving at speeds that could be as high as that of the blades in the kitchen blender and whirling around with it are electrically charged particles to nearly the speed of light. Fantastic, yes, but apparently in good accord with calculations and fitting well with these new observations.

Unfortunately, and much to my surprise, the initial reception of these ideas among my astronomical colleagues was not good. A conference on this new topic was organized by the NASA Goddard Institute for Space Studies in New York and the Belfer Graduate School of Science of Yeshiva University, for May 20 and 21, just 3 months after the first announcement of the discovery. The conference was entitled "First International Conference on Pulsars" and the organizing committee was headed by Professor A.G.W. Cameron of Harvard. Many speakers were invited, and some of the original discovery group came over from England. I had sent my paper to the organizers, prior to its publication in Nature, with the request for a brief time slot at the conference to present these considerations. Their response was, "Your suggestion is so outlandish that if we admit this for presentation to the conference, there would be no end to the number of other equally crazy suggestions that we would have to admit." I was not allowed to speak formally, though I got in a few words from the floor. Since by then I felt very sure of my explanation, I wanted to have this known, despite the views others were taking at that time.

Meanwhile John Maddox, the editor of Nature, obviously thought well of my model. The paper which was received in the offices of Nature on May 20 appeared in print on May 25, probably a record in publication speed for Nature.

The New York conference turned out to be a fiasco, dominated by a report of startling observations that turned out to be based entirely on equipment errors. The theoretical models that were discussed were almost all models of radially pulsating white dwarf stars. Both at that conference and in the many theoretical papers that were also published in Nature, the problem of the timing accuracy and of the great radiation intensity, which had been central to my considerations, were not discussed.

I did not have to wait very long for the confirmations of my theory. By October 17 of that year, the Australian Observatory at Molonglo (a branch of our Cornell-Sydney University Astronomy Center) reported the detection of a pulsar with 0.089 s repetition frequency, the shortest to date, and that in a location that was, to the detection accuracy, that of a supernova in the constellation of Vela that had exploded as recently as 20,000 years ago (as judged by the speed and size of the expanding cloud surrounding it). Since few recent supernova sites are known in our galaxy, this was clearly very suggestive, particularly as it also showed the higher pulse repetition frequency predicted for young neutron stars.

By November, Arecibo announced the clinching discovery. A graduate student of ours, Richard Lovelace, had prepared a computer program that would allow a careful search for pulses of much shorter periodicities to be made, even if the individual pulses were deep in the noise level of the receiver. This was applied first to a search in the position of the Crab Nebula, the most recent supernova site known precisely, having exploded in 1054 A.D. Success was immediate: they observed the fastest pulsar to date, 33 ms repetition frequency, in just that position. Within a day of that discovery, a slight but clearly measurable slowdown of about one part in 2,400 per year was also noted.

The National Radio Astronomy Observatory at Green Bank, West Virginia, had seen some individual short pulses of radio radiation from the direction of the Crab Nebula and had correctly assumed that these were from a pulsar in that location. But with an observation of individual pulses only, and without a regular repetition rate, they could not be certain.

The Crab Nebula had been an enigma since this diffuse nebula shone with a luminosity of about 100,000 times that of the sun, and yet one had never identified any energy source for this. I remembered this very clearly from my discussions with Walter Baade a few years earlier. We discovered then that the light was emitted along filaments, with a polarization at right angles to each filament. Energetic electrons gyrating in tangled magnetic fields would do that. But what produced this prodigious flow of electrons? The observed spin speed and slowdown, and the expected moment of inertia of a neutron star, allowed me to calculate right away the amount of energy that was being lost. Could this account for the energy radiated out from the luminous nebula?

I recall the exciting moment when I had sent my assistant to the library to look up the estimate of that luminosity. He came back with a figure that was remarkably close to the energy loss I had calculated, in fact, no doubt accidentally, within 20% of it. At this stage, there were no doubts left about the explanation, and I submitted this consideration in a second paper to Nature, received there on December 10.

A new and important era in astronomy had begun. An astronomical world of very high density and very high energy concentration had opened up. This was a world in which relativity theory was essential, not a minor correction to Newtonian gravitation. It was a world not only of strong radio pulses but of X-rays and high-energy particles. The fact that neutron stars existed, which represent a concentration of matter that is close to the relativistic configuration of a black hole, also meant that slightly more massive objects, which would indeed be black holes, could be expected to exist also. High-energy astrophysics and relativistic astrophysics had become a real subject of study (Fig. 8.4).

In the following years, many important discoveries were made in this field. Arecibo became by far the most important observatory for the observation of pulsars because most pulsars are weak sources and Arecibo continued to hold its position as the world's largest and most sensitive radio antenna. More than 300 pulsars are known at the time my writing of this memoir, most of them with higher pulse repetition frequencies than the first discovered ones, and almost all of them demonstrating gradual slowdown. The most remarkable observations were made by

Fig. 8.4 John Maddox, the editor of Nature, phones from London to tell Gold that his rotating neutron star model of pulsars has been accepted for publication on May 25, 1968

Professor Joseph Taylor of Princeton and his colleagues, who found that timing accuracies of many fast pulsars were comparable with the accuracies of the best atomic clocks in the world, after allowance was made for the mean observed slowdown rate in each case. At the present time, it is arguable whether atomic clocks or pulsars should be used where the most accurate timekeeping is required. The accuracy is of order 1 part in 10^{13}, which means that the timekeeping over a period of a year will not be in error by more than a few millionths of a second. Everything the pulsars do is in the realm of the superlative.

These timing accuracies also imply an unbelievably high accuracy with which the star keeps its shape. If a 20-km star expanded or shrank by one millionth of a centimeter, that is, 1/100th of the length an optical microscope can resolve, its frequency would change by an observable amount. We can be looking at an object hundreds of light-years away and confirm over the years that its size has stayed constant to 1/100th part of the wavelength of light.

The high timing accuracies of the pulsars, which Joe Taylor pursued so diligently, opened up new and powerful studies of general relativity. Taylor and colleagues, working at Arecibo, discovered a pulsar of very short period, and the subtle fluctuations in that period revealed many circumstances about it and its surroundings and have given us by now several of the most interesting confirmations of Einstein's general theory of relativity. The first thing seen was that the very fast, 59-ms, pulse period showed a small regular modulation with a period of 7.75 h, in just the way in which it would if the pulsar orbited another massive object in the elliptical orbit required by Kepler's laws of orbital motion. This, of course, would make a frequency shift (by the Doppler effect) to a lower frequency when the motion was away from us and to a higher frequency when it was toward us. But, of course, 7.75 h was a very short period for such an orbit, and it was clear that the other object, as yet unseen, was also a collapsed dense star; any ordinary star could not be orbited by another massive star at such close distance. From these variations, all in a single string of pulses, it was possible to deduce the inclination of the orbital ellipse to the line of sight, the real—not just the projected shape—of the ellipse, the masses of both stars, and the maximum speed of the pulsar in its orbit—400 km/s. Also the relativity effect, whereby the axis of the ellipse slews around slowly, could be observed accurately. This is the effect predicted by Einstein and first verified in the motion of the planet Mercury about the Sun.

Having settled this and having taken account of the slight and steady slowdown rate of the pulsar, Taylor then started to look at much smaller systematic variations in the pulse frequency, down to a precision of the order of one part in 10^{12}. A whole range of effects could be deciphered from them. He could observe another relativity effect, first predicted from Einstein's theory by Irwin Shapiro, namely, the extra delay that appears when a light ray (or a radio signal) passes close to a massive object. This effect had been observable in signals of spacecraft that were orbiting Mars, when they were thus carried to the other side of the Sun from us, so that their signals passed close to the Sun. It was a small but very accurately observed effect then, but here it was a large effect. The inclination of the orbit to the line of sight was such that the rays would pass close to the unseen massive object in every orbit, and an extra time delay could be measured every time.

Then, these extremely accurate observations led to the greatest prize in the relativity field: the first observation of gravitational radiation. Einstein had thought that gravitational radiation should exist, according to his theory. But he did not do any detailed studies to specify more closely the properties of this radiation. Bondi and his colleagues in London, at King's College, had worked on the problem in the 1950s and had concluded in detail what the theory predicted in this respect. It was clear that the effects would be unobservably small for any masses in our neighborhood. Fast motions of massive objects, and motions that deformed the gravitational field at a distance, were required. Not knowing what might be going on in other parts of the universe, experiments were being set up to detect such waves, if they happened to be present at greater strength than one had any reason to expect—a remarkably sensible attitude and an attitude that would have advanced astronomy enormously in earlier years when it had been the rule to look only for what one already knew to be there. Pulsars are indeed a case in point, for the equipment and the techniques for their discovery existed long before 1967, but no radio observer searched for fast time variations. Astronomical objects were thought to be so large that such variations could not exist. It was only the chance of using an instrument that was designed for another purpose where fast time variations were expected—

plasma effects in the interstellar gas—that led the young and inexperienced graduate student, Jocelyn Bell, to the discovery. Possibly her inexperience helped her, for otherwise, she might have discarded the observations as interference from a farmer next door; an astronomical source could not possibly make such fast signals. At any rate, by now, people were beginning to look for the unexpected, and Joe Weber, followed by others, had devised instruments to detect that jelly-wobble of the structure of Einstein's space–time that would constitute gravitational waves. Perhaps some enormously intense event somewhere would produce a disturbance large enough to be observable here. But so far, no such signals have been observed. Gravitational waves, like electromagnetic waves, would have to be propagated at the speed of light, as the logic of relativity theory is completely dependent on there being only one velocity for the propagation of any kind of signal through empty space. If there were two different ones, the most dreadful inconsistencies would arise.

Joe Taylor's approach had a much better chance of success. He was looking for effects of the emission of gravitational waves, not the reception. But those effects he could observe in the distant, best places, even if the energy radiated out there was much too small to produce anything observable here. The calculations by Bondi and colleagues (and others later) were quite explicit about the gravitational wave energy that should be radiated out from a close pair of rapidly orbiting massive stars such as Taylor was observing. That energy had to be supplied by the energy in the orbiting motion. Taking away energy from that would make the orbit shrink, the speed would go up, and the period of the orbit would become shorter at a rate precisely predictable now from the data of masses and speeds that were in hand.

Over a period of 15 years, Taylor followed the speeding up of the orbit. Year after year he could predict where the next data point would come, and he was never wrong. By now, there can be not the slightest doubt left that the pulsar's orbit is decaying at precisely the rate that the relativity calculations had predicted. The emission of gravitational waves had been proven; gravitational waves were real. From the point of view of theoretical physics, the proof of emission is just as valid as the proof of reception would have been. This discovery, and the refinement of techniques that went into it, must surely rank as one of the high points in the progress of observational astronomy and experimental physics. What else will be discovered from this new physics laboratory in the sky?

Chapter 9
NASA: The Love–Hate Relationship

The National Aeronautics and Space Administration (NASA) was set up by President Eisenhower in 1958, clearly in response to the Soviets' launching the first Earth-orbiting vehicle, the Sputnik. The propaganda value for the USSR had been enormous, especially since the Sputnik followed on the heels of a "missile crisis." This was the belief, probably false, that the USSR was far ahead in the construction of intercontinental missiles capable of carrying nuclear warheads.

NASA was set up to show that America was doing something about a perceived Soviet threat to American security. Unfortunately, many persons both inside and outside NASA thought that nonmilitary space activities were mainly to be a showpiece of the country's technological prowess. Manned flights were high on the priority list, and the progress from Sputnik and the early unmanned small spacecraft to manned flights was remarkably fast on both sides of the Iron Curtain. This seemed to be where the public interest lay. Gherman Titov, Yuri Gagarin, John Glenn, and Alan Shepard were the heroes of the day the world over. The scientific possibilities opened up by space flight took second place and the importance of space-borne telecommunications systems and of weather observation and prediction were not adequately recognized either by the public or by congress. NASA believed right from the start that its financial future lay in manned flights, no matter whether this was serving any ulterior purpose or not.

I recall that in the early NASA years, I tried, in publications and in speeches, to set the matter right. I explained that communication satellites would readily become an important business and their development costs would be amortized quickly. Even satellites that could distribute TV directly to the personal TV receivers seemed possible. This alone, I thought, could become the mainstay of NASA business. The interesting science that could be carried out with small unmanned spacecraft would surely tell us new things about the outermost atmosphere and the surrounding space. Astronomical observations without the obstructions of the atmosphere and observations of the Earth itself from outer space would answer many puzzles created by outbursts from the sun and the numerous phenomena, including the aurora and the large temporary changes in the Earth's magnetic field.

Thomas Gold *edited by* Simon Mitton, *Taking the Back off the Watch*,
Astrophysics and Space Science Library 381, DOI 10.1007/978-3-642-27588-3_9,

Getting to Know the Moon

My first direct contact with NASA arose while I was still at Harvard: I was appointed to be a member of NASA's first Science Advisory Committee, set up by Dr. Robert Jastrow, at that time, NASA's principal science advisor. One of the meetings that Jastrow organized was a discussion about the Moon, and it was arranged for one of the large television networks to screen this discussion nationwide. The Moon was of course a major topic of discussion since there was little question that unmanned spacecraft at least could be constructed to explore it. A small group was brought together consisting of Zdenek Kopal, a Czech-borne astronomer concerned with the Moon; Harold Urey, the discoverer of heavy hydrogen and a Nobel Laureate; and Gerard Kuiper, a major figure in American planetary astronomy; and I, then a professor of astronomy at Harvard. During this televised interview, I felt that Gerard Kuiper, known for his intolerance of any opinions other than his own, went out of his way to ridicule the notion that a fine rock powder covered much of the Moon. I was by then quite sure that my notions were correct because radar measurements of the Moon had been done in Britain, and these showed a very low radio reflectivity. This and other features of the radio echoes just could not be reconciled with a rocky surface of bombarded frozen lava. But for a surface composed of fine powder to a considerable depth, that was indeed to be expected. I explained this in the course of the debate, but it did not impress Kuiper, who was not well acquainted with the radio techniques. I had brought along a little transparent box filled with a dark fine rock powder, which I thought would closely match the lunar surface material in appearance and I meant to show this to my colleagues and the television camera. In a theatrical gesture for the TV cameras, Kuiper picked up this box, opened it, and blew the powder out before it could even be shown. It just made a dirty cloud over the table and made us all cough. He then said: "You can't really believe that such stuff would be on the Moon." His own belief had been that all the flat plains on the Moon were huge seas of frozen lava. Urey, on the other hand, took much interest in my arguments and invited me to give a colloquium on the subject in Chicago (Fig. 9.1).

The Van Allen Radiation Belts

One of the early and most successful space science enterprises were small Earth-orbiting satellites with some very simple equipment, organized by James Van Allen. Rather than going to NASA, Van Allen had sought out the help of the Army Ballistics Missile Program that had an establishment in Huntsville, Alabama. This group still contained a number of the German engineers that had worked on rocketry during the war and then had been brought over to the USA. The Army Ballistics group was indeed the first to launch a satellite into Earth's orbit after the Soviets. Van Allen, who is an extremely skillful experimenter, brought in

Fig. 9.1 A Cornell University public relations still of Gold explaining his ideas on the nature of lunar dust in a televised interview, 1965 (Cornell University Office of Public Information)

equipment designed to observe the dependence of cosmic ray bombardment on altitude. His main work had been in cosmic rays, and this seemed the natural next step for him, following the instrumentation that had previously been carried to the highest levels that balloons could reach. He now could reach the higher levels at which orbiting spacecraft fly. Moreover, the instrumentation required to measure the flux of the high-energy particles that are the cosmic rays was comparatively simple and could be adapted to the smallest spacecraft that could have radio communication to the ground.

The first such satellite that soared aloft gave an alarming result. Above a certain altitude, the particle counter stopped giving counts. This raised a question: Could the cosmic rays just stop at some height? What was known about the cosmic rays was the particle energies and their distribution in space. This would make such an explanation impossible. So what then was going on?

Van Allen soon hit on the right explanation for this. The particle counter had a time constant for recovery after making a count, and this meant that if the count rate were far higher than anyone had imagined, it would not be able to recover between counts, and thus it would give no counting signals at all. This would mean that the flux of particles that was present up there was quite different from the cosmic rays since there was no way that those could be confined to some limited orbits. Could there really be an enormously larger flux of particles at a height at which there was no atmosphere to stop them? Where would they come from?

Van Allen and several other scientists immediately pointed out that in the magnetic field of the Earth, there were stable orbits in which high-energy particles could be accommodated. On such orbits, charged energetic particles would spiral around the Earth's field lines and oscillate rapidly in latitude between the points where the converging fields would reflect them. Then, calculations showed that any such system of particles would precess slowly around the Earth in longitude. Soon, such orbits were calculated in detail and were found to have long-term stability. Once I understood that, I was surprised that no one had made such calculations before the discovery and predicted that this phenomenon was possible.

The well-known diminution of the magnetic field of the Earth during the main phase of a magnetic storm required a current ring around the Earth somewhere above the surface and probably above the atmosphere. This had been adequately discussed, but neither the origin nor the stability of such a ring ever had a thorough discussion.

I was in close touch with Van Allen in these exciting times as I had known him for many years as a cosmic ray experimenter. I spent many an evening in the basement rooms of the physics department in Iowa City discussing all the strange results that this experimentation was producing. All the mathematicians who had concerned themselves with this problem had come to the conclusion that such orbits of charged particles would be permanently stable. But then, of course, this introduced a question of how any particle could ever be deflected to go onto stable orbit. If it can't get out, it can't get in.

A sophisticated trick was suggested for the solution of this problem. Here's how the trick worked. Let's suppose high-energy particles that are not charged, such as neutrons, are sent out by the sun. Neutrons will spontaneously decay into protons and electrons. A certain fraction of those neutrons would penetrate some way into the domain of the stable orbits, undeflected by the field because they had no charge, and they might then perchance suffer the decay that turns neutrons into protons and electrons. This would be a way of putting charged particles into orbits that would then be regarded as closed and permanently locked in.

Although many people thought that this was a clever solution to this puzzling problem, I considered it quite inadequate on quantitative grounds. On the other hand, at that stage, I did not have an answer either. I wrote a paper for Nature describing these orbits and noting a few deductions one could make about them.

Naturally, the problem continued to interest me. I tried to puzzle out how particles could get into orbits from which they could not escape. Mathematical orbit calculations were all based on the Earth's field lines being anchored to the solid Earth. But this consideration was incorrect because the atmosphere is essentially an insulating layer, and a connection of the field lines through this atmosphere to the surface of the Earth does not tie any particular flux line out in space to a particular location on the surface. These outer regions could exchange positions and, with that, could exchange the particles that were trapped on them. Such interchanges would be very pronounced mainly at the times of solar outbursts and would thus allow particles from such outbursts to be transferred into trapped orbits.

Realizing now what an intricate system was fashioned by the magnetic field lines of the Earth above the atmosphere, I thought that there should be a special name for it. I proposed the word "magnetosphere," to denote that region around the Earth in which the Earth's magnetic field is the dominant quantity. I proposed this word in 1959 in an article entitled "Motions in the magnetosphere in the Earth." In the paper, I described the interchange motions that would allow energetic particles to be transferred between different orbits.

When I first introduced the new word magnetosphere, I was teased by some colleagues on the grounds that we don't need to introduce new words, and that in any case, the region was not spherical and other similar objections. However, within a few months, it was a standard word of scientific language. The Oxford English Dictionary still cites my paper as the first use of this noun in the scientific literature.

Washington Committee Work

In 1963, the White House decided that it should have more direct information about scientific subjects than it was receiving from NASA. A space panel was set up as an adjunct to the President's Science Advisory Committee, and I was invited to be a member. Two years later, I was appointed also to the major NASA committee, the Lunar and Planetary Missions Board, intended to guide NASA researches in these fields. It was a committee made up of scientists from outside NASA, mainly from universities and research institutes. In the course of this committee work, I got to know about the internal workings of NASA, of the relationship between NASA and the White House, and the various forces that seemed to be shaping the NASA program. The central conflict was whether manned space flight should be the ultimate objective or should instruments without humans do the scientific investigation at a much lower cost. In all the 30 years of manned space flight that I personally experienced, I have not heard of a single case where the presence of a human in the vehicle was of significant importance, except for the medical information of the effect of space flight on the human body. But if one did not need manned flight, one would not need that medical information. NASA, on the other hand, thought it could rely on the public to support the thrill of manned space flight, and this continued to be the NASA policy throughout my professional life (Fig. 9.2).

In the years between 1959 and 1969 (the year of the first Apollo landing), I was generally on very good terms with the NASA Space Science branch, and I received funds for experimental work, mainly related to the study of the Moon. In 1966, funds were made available by NASA for Cornell University to establish a building for Space Sciences and Astronomy. In 1967, this was completed, and the opening ceremony was attended by the deputy administrator of NASA, Mr. Seamens, who expressed NASA's great appreciation for Cornell's work. The Center for

Fig. 9.2 Tommy in his office at the Space Sciences Building, Cornell University, about 1970

Radiophysics and Space Research and the Department of Astronomy now had an excellent building with nice offices and a large amount of laboratory space.

Among the various research projects we had set up was one that I regarded of particular interest and significance to the lunar program of NASA. It was the detailed study of intensity and polarization of light scattered from various surfaces, and the comparison of those results with the reflecting properties of the Moon that had already been measured by the radar astronomers. Without a laboratory study, the astronomical information could not be used toward any prediction of the type of surface that a lunar landing spacecraft would encounter. With the help of Bruce Hapke, a good experimentalist, this program came to a very satisfactory conclusion. A stunningly precise match was obtained between one of our laboratory-created surfaces and the lunar observations. While we had tried many types of surface, none of the others came close. This was a match in both intensity and polarization for all angles of the incident and reflected light. The surface that matched was one composed of dark, fine-grained rock powder deposited by shaking it out of a fine sieve. The uppermost surface so created contained many peaks and low spots on a scale of millimeters. The small grains tend to stick to each other, and when they simmer down at random, they create that type of surface. Bruce Hapke gave it the name "fairy castles."

But this extreme fluffiness would not persist under more than a few centimeters. Several other lines of investigation, Earth-based radar and a variety of thermal measurements, also came to the conclusion that a fine powder must cover almost all the Moon. The radar information indicated the absence of any sharp transition to a dense rock at any depth less than 100 m.

All this fitted in well with my story that some surface creep must be taking place, which had removed large amounts of soil from the older mountains and turned the low areas into flat plains. Small grained dust could do this under electrical surface forces.

With all this evidence, I traveled to NASA headquarters in Washington, a sample of the powder and the graphs of lunar reflectance and that of our powder in hand. Homer Newell, the deputy administrator for space science, had collected a few engineers and also some prospective astronauts for this viewing. I had brought a microscope, and I assured them that this would be very close to what one would see almost anywhere on the lunar surface. Homer Newell saw the significance of the match, and he appreciated that grains of larger size would not make so close a match. Some of the others were not so sure, and they thought that there could be thousands of other and totally different surfaces that would have given the same result. Those who advised NASA on lunar geology were certainly skeptical of the whole story because they dug their heels in: They "knew" that the Moon was covered with frozen lava. My offer to construct a dust room for training the astronauts in the materials they would find on the Moon was turned down. Instead they were trained in brittle, sharp, angular rocks in lava beds and Meteor Crater, in northern Arizona.

Apollo

I was not really opposed to the manned space flight to the Moon, because this belonged to the realm of national scientific achievement, and the decisions for that have to come from the leading politicians. However, I expressed my own view that such a program could not be justified as one for science.

When Moon research by unmanned lunar landing vehicles got under way, one might have guessed that one of the first questions to be asked would be: Did the vehicle land on hard rock or on soft powder?

In 1966, the Soviet space program succeeded in landing Luna 9, a small unmanned spacecraft, on the Moon. The reports issued immediately stated that the transmitted pictures gave quite conclusive proof that there was absolutely no dust on the Moon and that this notion could now be laid to rest. This was based on the pictures of the surroundings of the vehicle, which showed sharp, angular features on a scale of a few centimeters, and the researchers had considered that only solid rocks could produce these. On the Earth, that might be true; there wind and water would have planed down such features if they were made of a powder. But if one takes baking flour and dumps it on a board, one finds that it makes equally sharp features. So these arguments did not impress me (Fig. 9.3).

If the surface had been one of dust, I suppose the researchers would have expected to see the track from the landing point to the place where it came to rest, no doubt some way away since it would have had a horizontal component of velocity at the rough landing. But there was a good reason why they would just not see that track: The object was a sphere constructed with four petals that opened after it had come to rest and then exposed a camera that had a panoramic view all the way around. If the sphere had come to rest in the ideal attitude, then the ground would have been seen in all directions. But obviously, it had not, since about half the panorama was just sky. It had come to rest in an altitude the designers had not intended. Why?

Fig. 9.3 Carrie, Tommy, and Lauren at Gold's 60th birthday celebration at Cornell University, 1980

The object had been designed as a sphere, so it could roll. Obviously, it was given a mass distribution so that the center of gravity was offset from the middle toward the point that was intended to be at the bottom when it came to rest. But if such an object rolls on a soft, dissipative surface such as one of powder, then, after rolling many revolutions, it will come to rest in the last turn when its center of gravity is on the way up, not on the way down. The panoramic view would then be one that looks down on the ground in front, and it would look upward in the back direction. That would just be the attitude where it could not see any track it had made. For that reason, it would not discover that the ground was soft and that there would have been a beautiful track with a smooth surface, as we now know from the later investigations. On a visit to the Soviet Union a year later, I spoke to some of the investigators concerned, and they told me that they were pretty sure that my explanation was correct. When the first US unmanned spacecraft landed, the US "Surveyor," it became quite clear that indeed a fine powder covered the surface since imprints of the finest detail could be seen on the surface made by the footpads of the craft in a bouncy landing. The scientific aspects of that program could have been done much more easily with unmanned craft with a remotely controlled roving vehicle, and a return spacecraft to bring samples of lunar dust or lunar rocks back to Earth. But, of course, after President Kennedy said "I believe that this nation should commit itself to achieving a goal, before this decade is out [1970], of landing a man on the Moon and returning him safely to Earth," NASA felt assured in their belief that manned flight was their ultimate objective. The White House Space Panel and the Lunar and Planetary Missions Board were not opposed to furthering a man on the Moon but were generally opposed to the outlook that only manned flights should define the future of the NASA program.

In our Cornell laboratories, we then proceeded with experiments to see what external influences could make the powder creep and generally to flow downhill. We could dismiss all of the solar radiation from the far ultraviolet through the visible into the infrared range. None of this did anything to our dust surfaces. Even X-rays were tried but showed no effect either. We then tried to bombard our dust surface in a vacuum chamber with a beam of electrons. The Moon dust receives

Fig. 9.4 Gold founded the Center for Radiophysics and Space Research in 1959

such a treatment from the Sun, so this seemed a realistic possibility. It was a fairly simple setup which Bruce Hapke fashioned, and the result was absolutely dramatic. In the range of 100–300 V of electron energy, the powder would hop and dance mostly only at millimeter heights but sometimes, at particular spots, to heights of a centimeter or more. If we fashioned a small step in one homogeneous powder, it would, on average, get smoothed out. All kinds of surface shapes were developed by this motion, and every new shape that had grown would again determine the next step. There was no settling down to a final condition. Even when the surface on an average was pretty flat, there would constantly be small-scale changes (Fig. 9.4).

Having gotten this far, we then attempted to understand this process in a greater range of circumstances. For instance, we made some mountains of the same material as the flat ground around them and some mountains of an equally fine powder but of a chemically different substance. The same powder mountains eventually just diffused themselves into the flat surface material, but the mountains that were made of another substance behaved very strangely. That material would also start out by moving down the hill, but then, when it came near to the junction line with the other material that made the flat plains, it piled up there. It continued to pile up until it made such a steep slope to the plains that pieces just broke off and fell down onto the material of the plains.

All these strange effects could be explained by the electrical charge distribution that a rain of electrons would produce and the differences in the charges from grain to grain that would produce electrical forces quite adequate to make grains jump. Of course, we were bombarding our sample with electrons at a much higher rate than that at which the Sun would bombard the Moon. However, we only looked at each experiment for perhaps 1 h, while the Moon would have suffered this treatment for four billion years (which is 3.5×10^{13} times as long). If we turned our electron beam down by some large factor, we would not see any movement. Nevertheless, the solar electron bombardment over the entire age of the Moon may well have been sufficient to cause the effects we have noted.

Several very puzzling features of the Moon had an explanation in these terms. The side of the Moon that faces the Earth looks very different from the side that never faces the Earth. The difference is that on the front, large areas of dark, flat, "mare" (Galileo's term) surface and flat areas inside old craters are very common. On the back, this is not so. Quite deep depressions exist that are not filled up. The mountains in general look less degraded than on the front, and few craters are filled with a flat deposit.

Relative to the Sun, the Moon rotates, and one might therefore have expected a similar bombardment. Yet all the differences seen, all statistically perfectly significant, indicate that the nearside of the Moon has suffered far more erosion and downhill transportation than the farside. It is only in relation to the Earth that there is a "nearside" and a "farside," but what effect could the Earth have that would cause more erosion on the nearside?

We asked ourselves if it was possible that the Earth influenced this electron bombardment. Gas from the outer atmosphere of the Sun is constantly propelled into space, a phenomenon called the solar wind. This gas is highly ionized, meaning that it consists mainly of electrons and of the nuclei of hydrogen and helium. At the expulsion from the Sun, the electron and the heavier nuclei are expelled to very similar velocities and therefore very different energies. The velocities of the electrons are then well below the 100 V level that we had required in our experiment. The direct impact of solar streams on the Moon would do very little. However, when the solar stream strikes the outer magnetic field of the Earth, a zone of turbulence is set up and that means that there is energy sharing between the nuclei and the electrons. In transit, the bulk of the energy was resident in the nuclei, but in this turbulence, the electrons gain a large fraction of that energy. This phenomenon has been known and quite well understood, and measurements with spacecraft have confirmed it. What it means for our purposes is that the Moon will receive bombardment with electrons in the energy range in which dust movement will be caused but mainly when it is in the drawn-out magnetic tail, drawn out in the direction away from the Sun. But this is the direction where the Moon will be for a few days around full Moon. The front face, or nearside, is therefore immersed in the magnetotail of the Earth and receives the energetic electrons. The back is never in such a position. The magnetotail of the Earth with its turbulent gas provided a perfectly good answer.

Another puzzling effect clearly seen was the nature of the junctions between mountains and the flat plains in which they stood. For the older craters and their mountains, we could perceive a loss of height compared with younger ones of the same crater diameter. But it was also evident that the slopes were generally much less steep. That by itself would not be surprising if surface material had suffered a preferential downhill motion faster on steep slopes than on gentler ones. But the junctions with the flat plains showed a shoulder region of much smaller inclination than the main mountainside, changing over to a much steeper inclination just above the junction with the flat deposit. This is seen on all or almost all the old mountains, and one therefore has to find a rational explanation for this. Quite large amounts of material are involved in these shoulders, amounting to several percent of the mass that is missing on the old craters compared with the younger ones of similar diameter.

We had noted the effect in laboratory experiments that at the junction line between two different substances, movement was greatly inhibited. We eventually understood why this was so, and we explained it as follows. If the charging of particles occurs in the region of transition so that particles of either type will be on the surface, one type will always charge positively and the other negatively, holding the average charge close to zero. The positive ones and the negative ones will therefore attract each other and get stuck in pairs or more likely in an array of long filaments, made up of alternatively positive and negative particles. Once such an array has been set up, none of the other electrical forces could break up this system, and it therefore becomes immobile.

This phenomenon will explain the shoulders that are seen on almost all mountain slopes, close to the transition to the plains. The partly eroded mountains are somewhat lighter in color and have indeed in general a different chemical composition from the flat plains. A particle making its way downhill will therefore come to a little strip close to the junction with the flat material and there it will get stuck. As other particles come down, they will similarly be immobilized and they will eventually pile so much material on the shoulder that a very steep slope from there to the flat surface will be created, so steep that material will just slide under its own gravitational force.

From Lunar orbiter photography, we now know that the final junction lines are extremely sharp. If you walked on the flat surface toward a mountain, you would find that from one step to the next you had changed from walking on flat ground to walking on extremely steep ground.

The Lunar Stereo Camera

I was keen to see some scientific information about the surface movements to come out of the Apollo program. I had persuaded my fellow members of the White House Space Panel that it might be important to see what the undisturbed surface of the Moon looked like. After all, the sample returned of lunar dust would not tell you that. I therefore proposed that a stereo camera be put together in the form of a walking stick with the operative camera at the bottom and looking down, and with a trigger on the handle. It would take stereo pictures with light from a flash gun. All the astronaut would have to do, as he walked across the surface, was to put the camera on the ground making contact with the shroud that extended out beyond the lenses to the focal distance to which the lenses had been permanently set. The light was fixed by the flash, the focal distance was fixed, and it would therefore be little trouble to take a lot of pictures while walking over the Moon. Features of the powder surface would be seen and could be compared with features we could create in the laboratory, and so this might check whether indeed electrostatic transport had taken place on the Moon. The White House panel agreed, and a committee was set up to direct the construction of such a camera. One of the members of this committee was Edwin Land, the founder of the Polaroid Corporation. Another was a senior scientist of the Perkin Elmer Corporation. I was appointed chairman of this committee. What we did was to draw up the specifications of the camera we wanted, such as the focal distance and type of film, the total number of frames that would be available, and we were then empowered to give this information to Kodak who would, with our guidance, design and construct the instrument.

It turned out to be a magnificent instrument and with the simple routine of picture taking, every picture turned out sharp and clear and with impressive stereo effect. The camera was one of the first instruments deployed on the lunar surface at the landing of Apollo 11. It was in use again later on the second Apollo flight, Apollo 12. A version was also on the ill-fated Apollo 13, and it was used on the Moon again on Apollo 14 (Figs. 9.5 and 9.6).

I had meetings with each group of astronauts before they flew the mission to tell them how we expected this camera to be used. I specified that it should be used on undisturbed ground, especially if it showed any features like small craters or any shapes that had no ready explanation. There were 110 stereo frames available on each film, and we expected the astronauts to take a large proportion of that number. After all, they had nothing to do but walk with this walking stick, hold it down briefly, and press the trigger. The pictures we got back were extremely good, but the trouble was that far too few pictures were taken. Instead of 110 we only got between 15 and 18 from the three missions that carried the camera.

Fig. 9.5 The stereoscopic Lunar Camera designed by Gold and used on Apollo missions

Fig. 9.6 Gold runs a careful eye over photographs taken by Lunar Orbiter 1966 – 1967

After the return of the Apollo 11 crew, I was at the session at which they were to report about their activities and their findings. I asked for a description of the appearance of the surface, and the answer was very surprising to me. The first response was that the surface looked "quite ordinary." I objected that I did not know what an "ordinary" lunar surface looked like. Then I got a more detailed report from them. There were some areas, they said, that looked randomly clumpy, but all the clumps were only powder. Then there was another type of surface, in some areas, that looked as if someone had gone over it with a garden rake. It had a lot of parallel lines that seemed to stretch as far as one could see and would even go down and up through little craters of which there were plenty. Then there was a third type of surface, which was described as looking like a beach, smoothed by a receding tide and then battered by a heavy rainstorm. Closely adjacent little pock marks were the result.

I asked whether any close-up pictures with the walking stick camera had been taken of these strange features, and I was very disappointed that the answer was no.

In response to other questions about the nature of the surface, they volunteered the information that it was nothing like the ground on Earth on which they had been trained, and that had been mainly fields of frozen lava. "We saw nothing that resembled this." The surface was powder and occasionally a few pebbles mostly at the bottom of small craters, pebbles that had glossy surfaces, sparkling in the sunlight. Even after the unmanned vehicles that had clearly demonstrated a surface of powder, the astronauts had still been given to believe that sharp, jagged frozen lava would be the surface, and I recall that the space suits were specifically designed for a great resistance to puncturing or cutting by sharp points or edges of broken lava rocks.

On the other missions on which this camera was carried, we did get a few pictures that showed the glistening little stones but also many pictures that showed the footprints several inches deep and demonstrating a precise molding by their footwear. What this meant was that wherever they had walked there were no pebbles embedded in the dust. It was all small particles and quite smooth and homogeneous.

Edwin Land had superb enlargements made of these pictures by a process to which he referred as "vectograph," where you would look with polarized glasses at a single transparency illuminated from the far side, and you would then see it all very clearly in three dimensions. These pictures had a wide distribution to many museums, both in the United States and in Europe.

What were the shiny little stones and what would they tell us? It was quite clear from the pictures that they had markings on them that looked as if a liquid had dribbled down over their surfaces. I presume that they were composed of rock material liquefied by impacts, in this case, quite small impacts, and then quickly frozen. It was still puzzling that they would be beautifully glistening since one would have thought that at least the small meteorites would have shot out clouds of dust as they hit and that the sum of this effect would have covered every other type of surface. We later came to understand from laboratory experiments that some materials can be "self-cleaning," in the sense that any dust grains deposited on them would tend to be removed by the action of electrical forces (Figs. 9.7 and 9.8).

On the whole, I was very pleased with the pictures I had, but disappointed that there were so few. After the third successful lunar landing, the use of the camera was discontinued. When I asked why, since it was such a small and convenient instrument and since many more pictures should have been taken, the answer came

Fig. 9.7 Tommy Gold and Carl Sagan in a lively debate, with Professor Yervant Terzian looking on, during Tommy's 60th birthday celebration at Cornell University

Fig. 9.8 Tommy's daughters Lucy, Tanya, Lauren, and Lindy

back that the astronauts had no interest in taking the camera again. So the decisions in scientific matters were made by the astronauts. I was not consulted, nor even informed of the decision.

Space Suits

Some months before the first Apollo flight, the White House Panel looked into the question of the design of the space suits: Were they safe? Did they allow sufficient mobility? Could an astronaut who had fallen over manage to get up by himself? The decision was that I should go to the Manned Space Flight Center in Houston and look into the subject. I did this, and my first request was to put on a space suit and have it pressurized so that I might judge the mobility it would afford. I was not impressed; the hands were very stiff; also, one could not bend over without pulling with considerable force on some externally mounted straps. I doubted very much that an astronaut who had fallen on his back in the soft powder of the Moon would be able to get up again. The astronauts in Houston demonstrated to me that they could, but this was on a hard floor, and they still believed that they would find this on the Moon.

In the conversations, it was reported that a company had designed a "constant volume" suit, in which the mobility was very much better. They showed me a movie of a man playing basketball, wearing such a suit. Sadly, they said, it is too expensive and too late to plan this for the Apollo program.

I was amazed to find that the idea of a constant volume suit was regarded as a novelty by the people who were controlling this development, and I discussed with them that of course anytime you have a change of volume when you moved some articulated piece of the suit, you would have a force, positive or negative, that would depend on the internal change of volume that this movement would cause. Of course, the internal volume has to stay constant if there should be no force opposing or assisting the movement of individual joints. I understood that it is not easy to design the suit that avoids changing volumes when you move in it, but I did not understand that there is any problem about understanding that a change in volume of the interior will either deliver or absorb energy and therefore either help or hinder the movement.

To my surprise, the people in charge of contracting for the design of space suits did not understand that, and moreover, were even unwilling to agree with it. In that frame of mind, they had merely seen the constant volume suit and admired the way that it introduced no unwanted forces in the movement of any of the joints. I would have thought that no one would be given the job of controlling the designs of space suits who had not understood this point completely.

I returned to Washington and reported all this to the committee, most of whose members could not believe that this simple point was not understood in Houston.

In thinking back on the complexity of the Apollo missions, of the life support systems, of the research equipment they would take to the Moon, and of all the many details of the work that had been spelled out to them, I was not too pleased to have found that in some quite basic and fundamental piece of equipment, NASA did not have personnel who understood adequately the physical principles of the equipment.

The Apollo program came to an abrupt end when there remained two vehicles, and the huge Saturn V rockets for those two final missions had already been produced. The explanation given was that there was not enough money to launch them. What really happened, as I saw it, was that the public interest in the last one or two Apollo flights was no longer what it had been for the earlier ones and that NASA would not spend its funds where not much public support could be gained. One of the lunar vehicles was converted to an Earth-orbiting craft called the Skylab, which would perhaps demonstrate something novel. I thought that this was the result of poor planning, to build these very expensive vehicles and then not to launch them for their intended purposes. I referred to this strategy as "buying a Rolls Royce and then not having the money to fill it up with gasoline."

Now 25 years later, it does not look as if that policy achieved anything remarkable. None of the Skylab results were of any consequence except to the flight of other Skylabs in the future. No purpose was served except perhaps to fly missions in the future that equally have no purpose. The enthusiasm of the public rather than realistic aims was the top priority of NASA.

This was still the situation as recently as the mid-1990s. NASA had become no longer competitive with several other countries for the launching communication and other utilitarian satellites. For unmanned instruments, modern rockets can do the same job far more cheaply than manned craft. Nevertheless, and I am sure in full knowledge of this, NASA recommended building the Space Shuttle, and in 1969, President Nixon decided to proceed with it.

Many members of the White House committee, including myself, opposed this strongly. We opposed it on the grounds that the financial predictions that had been made for the shuttle were absolutely absurd. It was claimed that it would cost as little as $50 for each pound to be put into Earth's orbit, which was at least an order of magnitude too small. It was claimed that 50 shuttle flights a year would take up many tons of instrumented hardware and place them in Earth orbit. So the United States became stuck with this shuttle program, which drained the country of more than $10 billion per year.

I had a remarkable encounter with James C. Fletcher, the Administrator of NASA, in 1971, in the midst of the Apollo program. With the strong encouragement of the White House panel, I suggested to Fletcher and Wernher von Braun, the German rocket engineer then employed by NASA, that we should have a meeting to discuss the pros and cons of manned flight versus instrumented flight with general-purpose remote control systems. Ed Purcell had paraphrased this as "put the eyes and hands of a man on the vehicle, but leave his stomach behind." What was meant by this is a device that now bears the name of teleoperator. The remote device consists of a television camera and a pair of arms and hands. The camera presents its view to the controller on the ground, and its movement is controlled by the operator's head movement with the television camera mounted in front of his eyes. The remote arms also move in detail according to the controller's arms. With such devices, one could operate any switches or one could make equipment changes, essentially all that a man in space could do.

Fletcher agreed to the meeting and its date was set for about 2 months ahead to make sure that all the major people in NASA who were concerned with manned flight could be present. I collected in these 2 months various statements from people who supported my viewpoint and also information about achievements to date with teleoperators, which was not very much. I also drew up a list of fields, unrelated to the space activity, in which teleoperators could be very useful, such as dangerous mines, nuclear reactors, and other places where one would need the eyes and hands of a person but where it would be unsafe to send a person. A development for which NASA would be the obvious prime mover would benefit many other fields. This technology has now been developed, not by NASA, but for entertainment, by the computer games industry.

The day before the meeting date, an article I had written on this topic appeared in the Sunday Magazine of the New York Times. Its title was "Machines, Not Men in Space," and in the article, I gave many of the arguments in favor of the teleoperator approach. Immediately after the appearance of that article, I had a call from Fletcher, and he shouted into the phone, obviously with great annoyance: "I have seen your article in The Times; I have canceled the meeting. The lines are drawn." I replied that he had long known my views on the subject and asked him if it made any difference that they now appeared in The Times. He just repeated: "The lines are drawn," and hung up.

I was invited to several congressional committee meetings to present my views and explain why the shuttle should not be built. One day before one of those meetings with a Senate committee under the chairmanship of Senator Walter Mondale, I was phoned by one deputy administrator of NASA who instructed me that I must cancel my appearance there for the next day, or else run the risk that NASA contracts to Cornell University would be stopped; not just the contracts that I had, but any and all NASA contracts held at Cornell! On the same day, the contract manager who dealt with my NASA contracts phoned me with a somewhat similar message. I deduced from this that the word had come from the top management, and it was their hope that I could be stopped. My response was that on no account would I succumb to such pressures.

The next day, when I went to the Senate committee, I privately told Senator Walter Mondale of these telephone conversations. He regarded this as completely illegal and wanted to take action against NASA in public in this matter. I was not in favor of this as I thought that it would seriously interrupt my work and that of my group at Cornell. Somehow we had to make our peace with NASA and legal action would not be the way.

Many years later, a historian went through ancient NASA files and he found a fascinating document. It was a letter headed "Eyes Only" (which means that it should never have gotten into any files but been destroyed after reading), and it was from George Low, the deputy administrator, to James Fletcher, the administrator, and dated April 17, 1973. One item runs as follows:

> Mondale request for Flax Report—So far Bill Anders has succeeded in not making this report available to Mondale. There seems to be no interest in the White House to do so. However, as of this writing, the letter to Mondale, informing him that he will not get the report, has not yet gone out.

> The item that concerns me runs as follows:
>
> Meeting with Tommy Gold—Tommy is interested in starting a teleoperator institute at Cornell and asked whether he could come to see me. His appointment was the afternoon after his participation in the donnybrook. I listened to his proposal (which incidentally is a good one) politely for about 20 min, and then told him that although I was very much interested in this kind of work, I did not think it would be appropriate for NASA to fund it at Cornell under his direction. I further told him that since his personal views were so totally adverse to what the Administration and NASA were trying to do in the space program, I didn't think that we could develop a useful relationship in any work that he might undertake for us. His immediate comment was that I was being narrow-minded and that just because he testified against us in the morning shouldn't keep us from funding him. In fact, he said were we to fund his project, the rest of the opposition would see that we are embracing not only manned flight but also unmanned flight through the use of teleoperators, and they might support us. He went on to say that with adequate funding he, too, might be able to support us. I tried to make it very clear to him that he was welcome to make any kind of adverse comments about the space program he desired, and that, in general, we certainly welcome debate of what we are doing, but we felt that this debate had best come from those not associated directly with the space program and not being funded by us. I had no objection to his testimony before Congress, but he should realize that being funded by the government and by NASA is a privilege and that it would make little sense for us to fund him as long as his views are what they now are. Tommy appeared to be quite thick-skinned, so I am not sure whether all of this sank in, but the message should have been clear.

Both these messages were highly questionable. Mondale had a perfect right to see the report for he had asked. So far as I was concerned, I knew that NASA as well as other funding agencies had the obligation to give out research funds strictly according to the quality of the proposal and not in any way influenced by other considerations. Now Low actually wrote that the proposal was a good one, and then he writes that it would make no sense to fund it because I was so opposed to the NASA plans. Fletcher had written, evidently on receipt of the letter, in his hand, against the paragraph that stated Low's reasons for turning down funding for the proposal: "great work." These remarks seemed to confirm the suspicion that I had before, that the NASA management had given the instructions to the officials concerned with my NASA contract to threaten me with withdrawal of NASA funds if I testified before congressional committees.

My testimony gave in some detail the reasons why I was so opposed to the shuttle program. The price tag for the construction of three vehicles was obviously quite unrealistic. The cost of each launch would be enormously more than NASA had stated. The total weight of instruments that were projected to be put into low Earth orbit with 50 flights in a year was about 20 times more than could be built or paid for. I also mentioned that a crash of one shuttle would interrupt the entire space program for a period of years, while a crash of an unmanned vehicle would not make much of an interruption. One senator who was present at a Senate committee to which I spoke later wrote me a letter congratulating me on my prophetic remarks.

In fact, they did not abruptly stop Cornell contracts as threatened, but instead made it harder and harder to get renewals. Altogether, the NASA organization seemed to turn against me. I was no longer invited to speak at conferences, publications of mine were no longer being cited, even in papers that were closely connected or even overlapped with what I had written.

Measurements of Dust Grains

We applied to NASA for samples of lunar dust returned by the Apollo missions. In particular, we could make three kinds of investigations in which we were interested and for which our laboratories at Cornell were adequate. One was simply a measurement of the distribution of grain sizes. Another was the detailed measurements of the electrical properties of the dust. A third type of analysis concerned the question of the reflectivity of the Moon. There are variations of the gray color of the Moon, but on the whole, it is extremely dark. Most ground up rocks would be much lighter. What was it that had caused the surface to be so dark, especially on the flat plains, and had made the mountainside a little lighter? We had previously proposed that this was the result of the solar wind, which would change the outermost few atomic layers of each grain. By driving hydrogen a small way into the surface, the material would be chemically reduced, and the proportion of iron atoms in the surface layer would thereby increase. This was to be expected, but we had also verified it with terrestrial rock powders.

Particles sizes turned out to be generally below 60 μm and were of quite irregular shape. The electrical measurements showed the material to be quite transparent to radio waves and to be attenuated to half the power level in dimensions of the order of 50–100 wavelengths. The dielectric constant values we found were quite in agreement with those the radar observers had given us. The penetration depth, which was so surprisingly large, would indeed have allowed us to see an echo from an interface with hard rock anywhere shallower than about 100 m. The radar observations that had been made at Arecibo by Dr. Tommy Thompson were at a wavelength of seven meters and showed no such effect anywhere, except on a small number of the youngest craters, specifically Tycho and Copernicus.

We extended our methods to search for the reasons of the darkness of the grains by adding what is called Auger analysis. This technique allows one to measure such quantities as the ratio of iron atoms to oxygen atoms in the outermost few atomic layers, and the results demonstrated that indeed this ratio was much greater in the outer layers than in the bulk of the material. The extremely dark Moon is due to the exposure history of the Moon to the solar wind.

Very little notice of this was taken at the lunar science conferences and in the scientific literature about the Moon since only a small proportion of the investigators had any interest in the dust. The radar observations were completely ignored and I don't think that any reference to them was made in any of the numerous lunar science conferences by anyone other than myself.

The Seismic Observations

Among the many different experiments performed by instruments landed with the Apollo missions, the seismic observations seemed to me the most interesting and remarkable. These observations with three seismograph instruments, placed at three different landing sites, demonstrated that the Moon was seismically extremely inactive, and most of the seismic signals received could be attributed to small meteorites striking the surface. I found this very difficult to combine with a picture of a Moon that had once been boiling and bubbling, a picture which I thought was just a hangover from the times when all craters were thought to have been due to volcanic action, not due to impacts. I had thought of the Moon all along as an object collected almost entirely from solid material in the solar system and generally too small to achieve the internal temperatures needed to make large quantities of liquid rock. Not much could happen and not much could move in such a body, and so it would be very quiet.

At a very high sensitivity of the seismic instruments, a few small genuine Moon quakes were observed. They came from approximately halfway from the surface to the center, and so some little activity has to be going on at that depth. The most interesting thing to note was that with those few genuine Moon quakes was associated the emission from the surface of a tiny amount of gas, recorded by a mass spectrograph standing at one of the landing sites. The Moon-quake-associated signal indicated frequently a mass of 16 atomic mass units. What atom or molecule could this represent? Atomic oxygen of mass 16 is out of the question since it would be chemically too reactive to go through the rocks. One had to look for a gas of 16 mass units; chemically, sufficiently inert; and found in abundance in planetary bodies. Only methane seems to satisfy these conditions. I presume that the Moon, like the Earth, contains a lot of carbon-bearing materials as well as water and that the products of this in the deep interior cause some gas to be developed, mainly methane and water. The water cannot come out freely because it freezes into hard ice at a depth of a few hundred meters.

The most important thing that the seismic investigations demonstrated was that the transmission of seismic waves was totally different from any such transmission on the Earth. On the Earth, there would be a sharp onset of the seismic signal at a time given by the velocity of sound in the ground and the distance between the source and the instrument. This would be followed by a short period of reverberation. On the Moon, in all cases that were recorded, the result was dramatically different. It took far longer for the first signal to arrive. It was not a sharp initial arrival but instead a gradually rising noise trace followed by an even more gradual decay. The whole of that signal occupied times of more than a half hour, while in similar circumstances anywhere on the Earth, the received signal would generally be over within one minute.

How could one possibly think that this signal had propagated through a thin layer of powder and then into solid, though heavily broken-up bedrock? If there was bedrock underneath, then, of course, the evidence of all the large craters would imply that it was heavily smashed, just as the young craters Tycho and Copernicus give us clear evidence of smashed-up hard rock. However smashed up this may be, the signal would still travel at a speed similar to the seismic speed on the Earth, and it would still deliver a sharp beginning pulse at the receiving end. Also, if the underlying rocks are badly smashed, a lot of the energy will be deflected in various directions including the downward one. The signal energy would therefore get weak very quickly, and the long reverberation time could not be explained.

I regarded this as an outstanding example and one of the few really valuable scientific contributions made by Apollo. No one else seemed to take any notice of it. My colleague Steve Soter and I wrote a paper in which we pointed out that this effect could not be explained by any transmission through solid rock but could be explained very well in a medium of dust, at least one kilometer deep, in which the compaction increases with depth. In such a situation, the sound velocity will gradually increase with depth, and almost all the sound energy will be refracted back to the surface. The surface of the Moon is of course uneven and so the reflections at the surface will again send waves down that again are refracted back to the surface. In this way, a reverberation of the ground would be generated first in the vicinity of the source and then, with innumerable refracting paths, it will reach the receiver. The reverberations will then continue until the energy is drained away from the region or until the absorption of the sound waves in the medium have used up all this energy.

The people who "knew" that there was solid bedrock just underneath continued to think so and seemed not to follow the arguments we presented. Somehow the broken bedrock must be producing this effect, they said, despite many examples on the Earth of very severely broken-up bedrock where nothing of this kind has ever been seen.

The small energy loss in the transmission on the Moon was due to two factors: One is that the powder in a vacuum tends to grow together at the contact points between the grains, and this makes such a powder an almost lossless transmitter of small amplitude sound waves. The other effect is that the sound waves only spread out in two dimensions, because of the refraction they suffer, due to the increase of compaction with depth and, with this, of the sound velocity. Very little of the power would go down into the body of the Moon.

This conclusion was one that fitted very well with the radar observations carried out at the Arecibo Observatory, which showed a steep falloff of the returned signal from the midpoint to the limb. If the signal was mainly due to the smooth surfaces and the worn down craters, then this could be understood. But if the signal penetrated to a layer of broken bedrock, then areas nearer the limb would return almost as much as those near the center. The difference between those two situations would be a factor in the returned intensity of more than a hundred and thus would far exceed any experimental uncertainty. A wavelength of seven meters had been used, and the laboratory measurements of lunar dust made clear that these waves would penetrate at least 200 m with little attenuation. The conclusion from both the seismic observations and the radar therefore tells us that a deep layer of powder covers almost all the Moon. That, in turn, means that all the powder represents small particles that fell in during the accumulation of the Moon, and not a ground-up bedrock.

Together with this goes the observation of core samples taken within one meter of the surface. Those showed individual layers of different reflectance and color, changing composition as quickly as in one centimeter from one layer to the next. A general destruction of bedrock by meteorites would leave a well-mixed surface. Similarly, trace element ratios showed different values by factors of 100 or more between the layers.

The Scientific Conclusions of the Apollo Flights

When the Apollo program came to an end, it seemed to me that everyone involved believed just what they had wanted to believe before the program started. I believed that a deep layer of dust with no sudden transitions to rock existed all over the Moon down to a depth that could in some areas be as much as two kilometers. The geologists who were concerned were delighted to find the composition of the surface material chemically similar to basalt, and so their case seemed to be proved. The Moon must have had huge lava flows since basalt on the Earth comes up in liquid form. So they had been sure in the first place that there would be nothing overlying the lava sheets, "not more than the dust on my piano." Then after the first few instrumented landings on the Moon, it was going to be three inches or perhaps one foot. The importance of this discussion of quantity lay in the following consideration. The hole that a grain-sized meteorite could make would be only a few times its own size. If we have three inches of dust lying on the Moon, all micrometeorites that fall in would either just contribute to this layer or be partly evaporated. If larger objects than the thickness of this layer had fallen in, they might excavate a crater down to the depth at which solid rock was thought to exist. That excavation would throw some chunks of rock around but will hardly add to the very fine powder that we now have. If this powder is now 100 m deep, then there would seem to be no way of accounting for it by the grinding up of any bedrock underneath. Such a layer would have to be explained as simply the amount of matter having fallen in grain-sized particles.

There is a class of meteorites (eucrites) that is quite similar to the lunar soil in chemical composition. And one might well think that the last accretion that built up the Moon consisted of some such material exploded off some other protoplanet. But this was not discussed: Basalt is basalt and it comes to the surface of a planet from underneath. So the fact that it was chemically similar to basalt seemed to indicate to them that it had welled up from the interior like basalt on the Earth seems to have done. Then, as I have said, the basalt could not be ground up into a powder to a depth of more than a few meters without the incoming material that does the grinding, adding more than the quantity it had ground up.

From a political viewpoint, the Apollo program had been a great success. For the scientific investigation of the Moon, its derivation, the nature of all the features that it showed, its internal structure, for all that, the mission and the organization was quite unsuitable. Remotely controlled vehicles, and the capability of a sample return back to the Earth, would have been much preferable because NASA could have done more flights, it would be a continuing program; instead, lunar research largely came to a halt. A research program of that kind would be very dependent on the ability to construct the next vehicle on the basis of the questions raised by the previous one. With small unmanned craft, this could reasonably be done. With the manned flights, it certainly could not.

The other great advantage that such a program for the Moon would have had would be the training and the experimentation necessary for doing a similar technique on Mars. In fact almost 30 years later, I still advocated doing such missions to the Moon to answer some of the major questions that the Apollo program did not answer but then to use the experience gained to proceed with similar techniques to Mars.

Space Manufacturing

NASA had dreamt for many years that there would be some important manufacturing process of something that required the absence of gravity. Organizations were set up to scout for such items. "Here we have a facility, please give us a reason for using it." Accurately round spheres for ball bearings were high on the list. The argument was that in the absence of gravity, one could just squeeze a little liquid metal in such a way that it would float while cooling and freezing. I had stressed over the years that surface tension, which would make droplets spherical, would be a weak force compared to the forces involved in crystal growth during cooling and that the final shape would result from this disorderly process. Nevertheless, such an experiment was in fact carried out on the Skylab mission, not with steel but with a metal of very low melting point. When the astronauts were asked how the experiment was going, the answer came back, "We sure have some funny shapes here."

The other space manufacturing exercise involved growing crystals of great purity. In the absence of gravity, convective motions in a liquid could be prevented, and these are sometimes a source of disturbance to crystal growth. The claim was made that the extremely pure crystals that could be manufactured in space would cure cancer and a number of other diseases, although no medical product was at hand that was seriously compromised by imperfection of crystal growth. It was also claimed that X-ray crystal analysis, such as that which led to the discovery of the construction of DNA, would greatly benefit. Max Perutz, one of the Nobel Laureates in the group, stated that for purposes of which he was aware, any large expenditure for space-grown crystals was not worthwhile.

I could judge the importance that NASA assigned to man in space for scientific purposes by the ease in which they gave away spaces in the shuttle and by the types of experiments that were carried out. What I regarded as joy rides were provided for an Arabian prince, for the senator from Utah, regrettably also for a young woman—a school teacher—and others who had no particular skills related to space flight. The senator from Utah tried to persuade me at length that he performed experiments which, he was sure, would wipe out cancer.

The Mars Delusion

In the 1990s, there were some in NASA who kept alive the notion that manned missions to Mars will be flown. They had not explained to the public that a manned mission to Mars would require a 2-year round trip time. It takes only 7 months to fly there, but then the astronaut has to wait a year before Mars and the Earth are again in suitable positions for the return flight. The total round trip time is thus of the order of 2 years. There are many reasons why such an operation is not practical. The first is that the expense is horrendous, and I cannot see any government making a commitment of that magnitude to so risky an operation. The bombardment by cosmic rays that the astronauts would suffer would be substantially more than they receive in the low Earth orbits, in which long-duration flights are now flown, where the magnetic field of the Earth holds off a good proportion of the cosmic rays that are in space. Shielding is not practicable, it would have to be several feet of concrete or the equivalent before it would even begin to reduce the cosmic ray damage. A small amount of shielding makes the problem worse, not better, because high-energy particles would create a shower of secondaries in that shielding that would still be sufficiently energetic to penetrate all that is inside the vehicle. And it would therefore make more damage than a single particle would have done. I thought this alone would rule out a manned mission to Mars.

On the Lunar and Planetary Missions Board, small unmanned research vehicles were certainly greatly favored over manned flights. Eventually, this NASA advisory committee was disbanded, I think, largely because NASA did not see why it should keep a committee that advised against their policy, which they were determined to follow.

In 1996, I wrote to the public relations office at NASA headquarters to ask whether they could send me a list of the major space missions, complete with dates since I wanted to write about them. What came back was a polite letter thanking me for my interest in space, and an inclusion of a pamphlet entitled "The First Thirty Years of Spaceflight." This pamphlet was entirely limited to the manned flights. All the flights of Shepard, of Glenn, and of the Apollo crews were there, but no mention was made of the glorious successes of the (unmanned) Mars landers or of the flights to Jupiter and Saturn.

My relations with NASA slowly came to an end. I received no more funding for researches at Cornell, and it was clear that I was regarded as the bad boy who almost killed the Shuttle program. It was approved by Congress by the narrowest of margins. The White House space committee also seemed to me quite unsuccessful in giving any direction to NASA, and one day in the time when President Nixon was embroiled in the Watergate scandal, I sent in a letter of resignation. According to the protocol, that letter was addressed to the science advisor to the president, Dr. Ed David, and the address on the envelope read "Dr. Ed. David, Science Advisor, The White House, Washington, DC." This letter came back to me unopened and with a big red stamp on it that said "ADDRESSEE UNKNOWN." The president's science advisor was unknown in the office he had directed; it turned out that he had handed in his resignation on the same day as I had done, so 2 or 3 days later when my letter arrived, the White House office did not know Dr. David anymore. I presume that this means that he left in a huff. It also meant that I had not yet resigned, and so I wrote a letter addressed to the appropriate office at the White House announcing my resignation. The only reply I received did not contain a word about thanking me for my long service on this committee or anything of that kind.

NASA Policy in the 1990s

NASA gradually came to the conclusion that unmanned comparatively small missions to the planets will be the best for planetary science. The Lunar and Planetary Missions Board had advocated that consistently, but NASA wanted to have more grandiose projects. NASA agreed to small and quickly repeated instrumented missions to other planets, but it still desired a huge project that would assure that it would have a very large budget. This major project is the International Space Station. A space station is essentially just living quarters in a low equatorial orbit, in which the astronauts can spend long periods of time. The intention was to have it permanently manned, implying that many shuttle flights a year would be needed to exchange crews and to provide the consumables. I think it was this aspect that made the program so attractive to NASA. NASA treated it as the next great step for the investigation of the cosmos, but very little was said about the proposed activities of the astronauts.

The public was led to believe that the space station was a necessary step for the assembly of that giant vehicle that would one day take a group of astronauts to Mars. But so long as we did not have a much faster means of propulsion than chemical rockets, there was no case to be made for preparing the assembly of a Mars vehicle in orbit. Although NASA did not propose any plan for such an expedition, presumably because it would be financially extremely unattractive and also very risky, it does not deny that it is planning for this and left the public and Congress largely believing that this was the purpose for which the Space Station was urgently required. If Congress did not believe that, NASA would not have stood much of a chance of being allocated the huge sums of money that were involved.

The greatest financial demand in relation to the space station will probably arise when safety measures are considered. If something goes wrong on the space station, we surely could not abandon the crews and leave them to their fate. Getting ready a rescue would take at least 1 month, and that surely is much too long a time.

If I look back over my last 30 years, I recognize that one might have thought 30 years ago that the presence of a man in a space vehicle might be an advantage. But then in those same 30 years, the advances in instrumentation have been enormous. The same instruments that can now be made with a weight of 10 g might have required 10 kg then. The reliability of modern electronics seems to exceed the reliability of even the most dedicated and trained individuals. The future for the investigation of our solar system will go more and more in the direction in which technology is progressing, and not in the direction of returning to the clumsy, dangerous, and unimportant manned flight.

Chapter 10
Origin of Petroleum on Earth

The Debate About the Origin of Petroleum

Although I had a long and sometimes fruitful association with NASA, and several people on my staff worked under NASA contracts, it became clear to me that this alliance had to come to an end. I decided to return to my longtime occupation of tackling outstanding problems in various fields of science. Among those, earth sciences were the most interesting to me at that time, and I became fascinated by the origin of petroleum, which is an important subject for many branches of science, such as geology and geochemistry, but also for technology and economics. I thought that it would have received the intense interest of scientists concerned with these fields. But in the developed countries, there is hardly any debate in this field; the majority opinion has declared it as solved. Biological debris buried in the sediments is supposed to decay into oil in the long course of time, and this oil is then supposed to become concentrated in the pore spaces of sedimentary rocks. The search for oil is conducted with this biogenic origin theory in mind, so that the presence or absence of biological material in the sediments was regarded as a key item. When I entered this field, very few boreholes were drilled into the basement rocks that had frozen from a melt, since there could be no biological sediments in those, and it was thought that they could not contain any oil. This belief then resulted, inevitably, in almost all oil being produced from sediments.

The alternative viewpoint was that hydrocarbons were a component of the materials that formed the Earth, and with increasing internal heat, liquids and gases were liberated, which forced their way upward and then may have been arrested temporarily in the porous rocks at depths that our drills can reach and from which we then derive commercial petroleum. A very large amount of scientific literature is devoted to this theory of the primordial origin of petroleum (abiogenic origin), but it is mainly in the Russian language, and this seems to have made it largely unavailable in the West.

The theory that biological materials formed the basis of hydrocarbons on the Earth first arose in the middle of the last century. At that time, there seemed to be

Thomas Gold *edited by* Simon Mitton, *Taking the Back off the Watch*,

Astrophysics and Space Science Library 381, DOI 10.1007/978-3-642-27588-3_10,

quite compelling reasons for it. The chemists had long called all compounds of unoxidized carbon "organic compounds," since they had not been able to generate any of them without the intervention of biology. There was also a reluctance to an outlook that biological processes were of the same kind as those in inanimate matter. By 1830, one "organic" compound, urea, had been created artificially, but this was not enough to change the outlook. To this day, all of chemistry is divided into organic and inorganic chemistry. Plants and animals contain unoxidized carbon, and they provided the only such substance that the chemists could find on the surface of the Earth. But the reason for this is simple: unoxidized carbon compounds would not be around for a long time in the presence of our atmosphere which contains one of the most aggressive chemicals known: oxygen. This resulted in the belief that only biology could currently create such compounds, and hence, their designation as "organic." The oil that was found to come out of the ground belonged to the domain of organic chemistry, and it was readily assumed that it had been produced by biological, "organic" processes.

This misleading nomenclature still affects the thinking of many people. When, for example, it was announced that "organic" compounds were found on dust grains in space, the letters I received on this topic made clear that many newspaper readers thought this proved the presence of biology in interplanetary space. When, many years ago, the gas methane, an "organic" compound, was found by spectroscopy to be present in large amounts on the planet Jupiter, the leading scientific journal, Nature, even published an article in which it was claimed that this implied that life must have existed on Jupiter. The chemists understand the situation well, and since they are not confused by this terminology, they see no reason to change it. They know full well that the term "organic" is no longer intended to have any implication that biology had anything to do with the origin of any particular "organic" molecule. The division of chemistry into the two branches is a good and useful one for them. But since the terminology still creates confusion among others, the term "organic chemistry" should long have been replaced by a term not carrying the same implication.

But this terminology was by no means the only cause for the belief in the biological origin of petroleum. Coal had long been regarded as deriving from plant material buried in swamps. There were many fossils in brown coal (lignite) that clearly suggested this, but bituminous (black) coal was regarded as having derived from lignite during long periods of burial at high temperature and pressure. Thus all types of coal were regarded as having a common origin, although black coals have very few fossils, and they are chemically of quite different composition. If the origin of bituminous coal seemed settled, it was not a large step to think of oil, a chemically very similar substance, as having a similar origin, though perhaps requiring a different type of biological debris and a different set of conditions of burial for its production.

A biological origin of oil seemed to be the only possible explanation at a time when it was widely believed that the Earth had formed as a body that was initially very hot. Some would have it gaseous; others thought that it formed as a ball of molten rock. If that had been the case, then any hydrocarbons that might have been present in the original mix would not have survived. Fluids of lesser density than rock would have been driven to the surface and more than enough oxygen would have been available from the magma to oxidize all that carbon. Whatever had reached the surface would have dissociated at the high temperature, and any hydrogen from it would have escaped into space.

Why did people think that the Earth had formed as a very hot body? I still remember learning at school, around 1935, that we were standing on a crust of a liquid Earth and that, in proportion, the crust was as thin as the skin of an apple. The reason for this belief was that all the bedrocks under the sediments were seen to be materials that had clearly frozen from a melt. It seemed a straightforward inference that the whole Earth had once been molten. The astronomers opposed the hot ball of gas hypothesis, just on the grounds that the small gravitational force of the Earth would not hold such a ball together, but they had no real argument against the ball of molten rock, even if they had no explanation of how it could have formed. The geologists, on the other hand, thought they had a strong argument for the hot origin of the Earth. (Today it is believed that all the frozen basement rocks have resulted from a partial melting of the mostly solid interior and that this melt oozed up in many locations and flooded over the surface in an early period of the Earth's history.)

Within the theory that the Earth had formed as a hot molten ball, the geologists thought that they had no option but to invent an explanation for petroleum that involved processes which could create hydrocarbons after a cool solid crust had formed. Carbon could only be expected to have been present in oxidized form on such a planet, and therefore, a source of energy would be required to turn this into an unoxidized form, to turn an ash into a fuel. Sunlight and the process of photosynthesis in plants seemed the only possible process that could be suggested, and this was readily accepted as the solution. Plants all around were turning solar energy into chemical energy, by the dissociation of oxidized carbon derived from the carbon dioxide of the atmosphere into "organic" carbon. It must have seemed at that time that this was the obvious explanation for all "organic" carbon on the Earth, including coal, oil, and natural gas.

But still, even then, there were dissenting voices. The great Russian chemist, Mendeleyev, had concerned himself with the question of the origin of petroleum. He recognized a number of features in the occurrence of petroleum that seemed to him to be incompatible with a derivation from biological debris. He noted the high hydrogen content of petroleum which he could not explain on that basis, and in particular, he saw the large-scale geographical patterns of hydrocarbon occurrence over regions within which there were quite diverse geological features and no coherent pattern of biological content. His conclusion was that "petroleum comes from the depth of the Earth, and it is only there that we must search for its origin."

Origin of the Carbon on the Earth

The surface and surface sediments on the Earth contain approximately 100 times as much carbon as would have been derived from the grinding up of the basement rocks that contributed to the sediments. The surface is enormously enriched in carbon, and this needs an explanation. This carbon is 80% in the form of carbonate rocks (oxidized carbon compounds) and 20% in unoxidized form. If the quantities are expressed as the mass of the element carbon per square centimeter of total Earth surface area (land and oceans), the estimate is about 20 kg. What could be the material that the Earth contains, which delivered all this carbon to the surface?

Meteorites are samples of leftover materials from the formation of the planets. Although they may not be representative of the quantities of the different types that made up the Earth, they appear to represent at least samples of all the major components. Only one type, the carbonaceous chondrites, contains significant amounts of carbon, and they contain it mainly in unoxidized form, a substantial fraction in the form of solid, heavy hydrocarbons. This material, when heated under pressure, as it would be in the interior of the Earth, would indeed release hydrocarbon fluids, leaving behind deposits of solid carbon. The carbonaceous chondrite material may be a major component of the Earth at its formation, and in that case, the surface carbon supply would have originated mainly in the form of hydrocarbons. Very little of the carbon in meteorites is in oxidized form, so a supply from an oxidized hydrocarbon source seems improbable. Even so, oxidation in the heavily oxidizing rocks or in the oxygen of the air would quickly have turned much of it into CO_2.

Meanwhile, we have seen different compounds of unoxidized carbon in huge total quantities in many places in the universe: giant molecular clouds are the largest known reservoirs of hydrocarbons. Brown dwarf stars that are cool enough to maintain methane in their atmospheres are seen to have the methane spectrum as the most prominent feature. More locally, we have seen hydrocarbons in the spectrum of the atmospheres not only of Jupiter but also of Saturn, Uranus, Neptune, and in particularly large amounts and great variety of molecular types, on Titan, Saturn's largest moon. Titan is the only moon in the solar system that has a substantial atmosphere, and there methane (CH_4) and ethane (C_2H_6) are abundant enough to be the major cloud-forming substances. A whole range of other hydrocarbon compounds similar to those in terrestrial petroleum have also been identified there. Lakes of liquid methane are present on the surface, where hydrocarbons seem to play a role quite similar to that of water in our atmosphere. Comets also puff out hydrocarbons, and the extremely dark surface of the core of comet Halley, seen by a spacecraft, suggests a surface of tar, which is a mix of solid or almost solid hydrocarbons.

It seems clear to me that of all the "organic" compounds in the universe, those made by biology, are the minutest fraction. This will be true even if there are lots of planets supporting biology around myriads of other stars. All of such biology could not come close to matching the mass of compounds of unoxidized carbon in the gas clouds of the galaxy.

One might think that the entire set of arguments about a biological origin would have come under fresh scrutiny once it was realized that "organic" molecules can readily be produced without biology; once it was known that the Earth had formed from cold, solid materials that could certainly have contained hydrocarbons just as the meteorites and the interplanetary dust that falls onto the Earth today, and when it was found that hydrocarbons are plentiful on other planetary bodies and in interstellar space. At that stage, it surely must have seemed very strange to argue that the other planetary bodies have hydrocarbons of nonbiological origin, as does interstellar space, but just the Earth, which also owns hydrocarbons, obtained them from a different source, one that is available only on our planet, namely, biology.

It appears that some geologists still take seriously what was taught a century ago. This is one example which shows the bad effects of this compartmentalization of modern science to which I have referred earlier. I now find it hard to see why there is still a debate, and why still the majority of investigators believe in the bio-origin theory.

I have identified what seems to me to be a general rule: when after a long time and with the lifelong effort of a large number of people, the definitive proof of their long-held viewpoint is still not available, and passionate belief takes the place of the established scientific method; the greater the importance of the subject, conceptually or economically, the more intense the passion.

Association of Hydrocarbons with Helium

The simplest and clearest case for a deep, nonbiological source of petroleum is its frequent, almost universal association with the noble gas helium. Helium is an element, and no chemical process can create it. Since it has virtually no chemical interactions with any material, no chemical process, biological or nonbiological, can cause it to be gathered up from a low concentration and brought to a higher one. Helium exists on the Earth from two sources: the radioactive decay of uranium and thorium produces it, but also some helium, possibly a large amount but a small fraction only of what is recovered now, was incorporated in the Earth at its formation. This was probably trapped in the solids that accumulated and built up the Earth. In outer space, helium is the second most abundant element after hydrogen; in our atmosphere, it is present only at the concentration of 5 atoms per million molecules of air. In many oil and gas fields, it is found at a concentration of 1 per hundred gas molecules, concentrated by a factor of more than 1,000 relative to the air.

The helium–hydrocarbon association is so strong that in the commercial search for hydrocarbons, helium sniffing on the surface was found to be useful. Very sensitive helium detectors exist, and they were first employed for the detection of uranium deposits underground, which were thought to be major sources of helium. That search was not very successful. But it was soon found that helium sniffing was indeed successful for the detection of oil and gas fields.

What could be the reason for all that? It clearly was a puzzle, but although hundreds of papers had been written describing these effects, very few investigators discussed what the explanation might be. Nothing other than a mechanical pumping action could concentrate helium into hydrocarbon reservoirs. Since we cannot think of such actions occurring in the horizontal plane, directed to the hydrocarbon reservoirs, we must consider an ascent from deeper levels.

Any gases or liquids that come up from deeper levels would pump up traces of any other gas they encountered on the way. Helium would be in the rocks, produced by the radioactive elements, but in so low a concentration that it would just represent individual atoms caught here and there in the rock crystals. These could not make their way up on their own. But if there are pore spaces and cracks held open by the pressure of other more abundant fluids which are moving upward, then the helium will be pumped upward with them. On the whole, it will be true that the longer the pathway that the carrier fluid has traveled, the more helium it will have swept up. Of course the content of uranium and thorium will be regionally variable, but not by such a large factor. Whether methane, for example, had traveled 1 or 300 km would almost certainly have a much larger effect on the amount of helium carried, than any variation of the mineral concentration along these paths. So this solves the puzzle of the association of an inert gas with other gases: the helium content is mainly an indicator of the depth from which the other fluids have come. If hydrocarbon gases have brought up the most helium, they must be the most abundant gases that have streamed up from deep levels. Nitrogen has frequently come from deeper levels still and then brings up a larger proportion of helium. But as nitrogen is much less abundant than methane, the hydrocarbons bring up the largest total amount.

So here, we have the conflict: the helium content of petroleum can only be explained by an origin far below the biological sediments. On the other hand, the biological molecules in petroleum—the so-called biomarkers—seemed to demand an origin from biological debris. The two arguments appeared to be in direct conflict with each other. Which one is wrong?

Microbial Life

The only way out of this dilemma that I could see was to suppose that there is an immense amount of microbiology in the subsurface, down to at least all the depths from which petroleum is extracted, and that all petroleum is therefore heavily contaminated with molecules derived from biology, while not being itself of biological origin. Estimates of the quantity by mass or volume, of the biological material necessary to produce the observed "contamination" of all oils, turned out to be enormous. There must be a whole domain of life there, of which we had been unaware. Actually, it cannot be too strongly emphasized that petroleum does not present the composition picture expected from modified biogenic products, and all the arguments from the constituents of ancient oils fit equally well, or better, with the conception of a primordial hydrocarbon mixture to which bioproducts have been added.

I began to ask myself if all the known biomarker molecules in oils could be produced by microbes. A few of these molecules occur also in meteorites, and they may not be of biological origin there. Some of these can be formed without the intervention of biology and are therefore not to be regarded as certifying a biological origin of the oil in which they are found. Among those, there is one type called "porphyrins" that is of special significance. These are complex molecules made up of carbon, hydrogen, nitrogen, and a metal atom. They are common in biological materials, deriving from chlorophyll in plants and from hemoglobin in the blood of animals. The chlorophyll molecule has magnesium as its metal atom, and the hemoglobin molecule has iron. The porphyrins in oil have nickel and vanadium. Not a single case is known of a magnesium or an iron porphyrin being found in any of hundreds of samples of petroleum from all parts of the world. The porphyrin story then argues just in the opposite direction to what had been intended: molecules from chlorophyll or hemoglobin must be very rare or absent in petroleum, and the porphyrins that are in petroleum are not due to biology. They would still require a hydrogen and carbon environment to form and would therefore be associated with petroleum.

Of the genuine biomarker molecules, many do occur in higher plants and animals, but so far as we know now, all occur also in microbes. Some are only produced by microbes.

The second major question is whether there really is this vast amount of deep microbiology down in the pores of the rocks that would be needed to contaminate all oils with such molecules. What supplies of food did it have?

Chemical energy is available in any location where a stream of a fluid percolates through the pores of the rock, if that stream is not chemically equilibrated with the material of the rock. This would be a common situation on a planetary body that had accreted from a mix of cold solids and had heated up subsequently. The volatile components would become mobile as gases or liquids, and if they are less dense than the rock, buoyancy forces would propel them upward. In the different rocks through which they might travel, there will be opportunities for reactions that supply energy. Reactions that run by themselves in these circumstances are of no use to biology; for example, a substance that will spontaneously combust in the presence of the surrounding oxygen cannot feed microbes. But there are many chemical reactions that require overcoming a threshold in energy; we all know this from having to set a lit match to a combustible material, to make it burn. Once it is hot enough, it will continue to burn. It is in such circumstances that microbes can get into the act. If they have the right kind of catalysts or enzymes (much the same thing), they can make the reaction go without the high temperature of the match.

In a rock flooded with hydrocarbon gases and liquids, there are certainly many possibilities of this kind. At shallow levels, hydrocarbon-oxidizing bacteria can use atmospheric oxygen dissolved in groundwater. At deeper levels where no atmospheric oxygen is available, it may be obtainable with a small expenditure of energy from oxides of iron or of sulfur (and perhaps other substances), and after spending the energy of liberating this oxygen, the microbes can then get more energy back by using this oxygen to oxidize the hydrocarbons. These would be processes that cannot run by themselves, but need the subtle intervention that bacteria can provide. Other reactions that deliver energy will also be possible, such as reactions involving hydrogen or hydrogen sulfide (the stinky stuff) and possibly many other gases and liquids. The Earth has an immense supply of chemical energy, and bacteria adapt to almost any situation in which they can feed on that.

Some oil comes from depths at which the temperature is in excess of 100°C, and nevertheless, it has some biologically produced molecules in it, but in much smaller quantity than shallower, cooler oils. What then is the real temperature limit for any form of microbiology? In the oceanic abyss, there are extremophiles that tolerate temperatures as high as 130°C, but only in circumstances where the pressure is high enough to raise the boiling point of water above that value.

Could microbial life have found its way down to great depths? This is not a problem. Just the growth of a bacterial mass along the pore spaces of the rocks would be fast enough to take them to the depths of the deepest oil wells in just a few thousand years, a very small fraction of the time available. They would of course only grow along pathways that supplied them with at least a minimum of nutrient. But in petroleum-bearing regions, one can always find some methane and other hydrocarbons at all levels above the main petroleum reservoir.

So I concluded that the dilemma was solved: helium can be associated with hydrocarbons, and at the same time, all the known petroleum biomarkers can be present. But the way out of the dilemma entailed a new and quite drastic assumption: a huge amount of microbial life had to exist at depth in the rocks.

The Deep, Hot Biosphere

I was greatly encouraged that this viewpoint was correct, by the discovery of volcanic cracks in the ocean floors with living material in and above them. These cracks stretched over thousands of miles and are now known in all major oceans. When they were first discovered with the aid of small submarines (submersibles) around 1979, observers saw a whole wealth of biology surrounding them, containing many species that had not been seen before. It became clear that all this biology was dependent on fluids that came out of these fissures, often very hot, and that brought up some substances that could provide chemical energy in combination with the local rocks or with components of the seawater. The large life-forms that had been seen all derived their food as the second step in the food chain. The first step, utilizing the diverse chemical energy supplies that happened to be provided, involved microorganisms that evidently were at ease in these remarkable living conditions, involving in many cases—but not in all—quite high temperatures. In the mix of different fluids issuing from the cracks were hydrocarbons.

It appeared that all this biology was dependent only on the chemical energy that the Earth supplied and was quite independent of the main energy source of surface life, the Sun and the photosynthesis of plants. It is of course pitch dark down there, and even the temperature is dominated by the heat coming from the Earth. The Sun and atmosphere have a minor influence only. If the Sun turned off, this life down there would probably continue, although a lowered temperature would force it to go a little deeper.

Why would such a microbiology not flourish also in continental regions where similar nutrients were coming up? The large life-forms seen in the ocean vents could not be there because pore spaces in the rocks did not provide enough space. But for the microbes, there would be plenty of space. Was there any reason to think that such life would not have expanded to the full extent the chemical energy supply could support? In that case, the total quantities of living material and its accumulated debris would be very large by any standards.

One estimate of these quantities can be made by the study of a set of molecules of bacterial origin. Professor Ourisson and colleagues in France have measured the abundance of a class of molecules that derive from the cell walls of bacteria. They found an astonishingly large quantity of these to be present in all the numerous crude oils they examined and also the same set in bituminous coal. On the basis of these observations, one could well derive estimates of the total mass of biological material existing at depth, which would be greater than all the mass of the surface life. Professor Ourisson considered this to prove the biological origin of oil and bituminous coal, all deriving mainly from bacterial life; yet in that view, the bacterial life would still have been dependent on the photosynthesis of plants and therefore on surface conditions.

I placed just the opposite conclusion on this. These molecules showed that bacterial life had flourished in all locations in which deposits of oil and bituminous coal had formed, and this had given them their biological imprint—just as Robert Robinson had said: "A primordial hydrocarbon mixture to which bioproducts have been added."

Using the estimates based on Ourisson's findings (and on other data also), one might well conclude that the deep microbiology and its products could be comparable in mass with that of all surface life; there may be an immense domain of life deep below the surface that we had so far failed to recognize. The surface life is restricted essentially to a few meters around the two dimensions of the surface. The deep life will be restricted to a pore space volume of only a few percent of the volume of the rock, but the third dimension, the depth, may stretch over 10 km or more and could harbor more than all surface life. Life based on solar energy and photosynthesis may be only one branch and possibly just an outgrowth of subterranean life. This suggestion, first derived from the need to solve the puzzle of the association of hydrocarbons with biomarkers and with helium, now opened up a whole new field. I believe that it will be a large item in the study of biology, including the origin of life, and of a large range of geochemistry.

I have written a much fuller account of this hypothesis in my book *The Deep Hot Biosphere*, published by Springer in 1999. My premise is that deep within the Earth's crust, there exists a second biosphere, composed of very primitive heat-loving bacteria and containing perhaps more living matter than is present on the Earth's entire surface. In the book, I join the deep hot biosphere argument to another, perhaps even more controversial theory for which I have marshaled the evidence: that so-called fossil fuels originate not from compressed biological matter at all but from deep within the Earth, present there since the planet's formation, long before our oxygen-rich surface biosphere came into existence. As I will explain in Chap. 11 of the book you are reading, the deep hot biosphere and deep-earth gas theories shed light on the nature of earthquakes; they suggest that reservoirs of petroleum and certain metal ores are much vaster (though not necessarily more accessible) than generally claimed, and they help to answer two of the most profound mysteries of the biological sciences: the origins of life on Earth and the prospects of extraterrestrial life.

The Origin of Diamonds

Diamonds tell us quite a detailed story about the physical and chemical conditions in the Earth, below about 150 km, and this information bears on the origin of petroleum. Chemical theory and the experience in making artificial diamonds show that high pressures of the order of 45 kbar are needed to produce this dense crystal. Such pressures are found in the Earth only at a depth of 150 km or more, and it is somewhere at such depths that natural diamonds must have been formed. The temperature there exceeds 1,000°C.

The geologic situation in which diamonds occur shows that unusual and gigantic eruptions were involved. Although many diamonds are found dispersed in river gravels, the only concentrated deposits are in the rare structures called kimberlite pipes. These are deep, vertical shafts, usually filled with a mixture of rock types, including the diamond-bearing rock called kimberlite.

The pressure and temperature at a depth of 150 or 200 km are in the right range for carbon to crystallize as diamond. But how did the carbon become concentrated? The two types of fluids that one may consider for the concentration process are carbon dioxide and methane, the latter possibly associated with heavier hydrocarbon molecules also. Tiny pore spaces in diamonds have been analyzed and found to contain small amounts of highly compressed gases, among which the carbon-containing ones were both carbon dioxide and methane. It is clear, therefore, that not only unoxidized carbon, namely the diamond itself, but also methane can exist down there. Thermodynamic calculations have shown that both these gases are stable in the upper mantle at diamond-forming depth, and either could be responsible. The following indications would seem to favor methane. Methane is generally much more abundant in the crust than CO_2 and appears to be streaming up from deeper levels. Secondly, of the gases contained in diamonds, nitrogen is by far the most abundant. One has to judge that nitrogen had something to do with the deposition of the diamond.

We see that the evidence from the diamonds is simple and clear. Unoxidized carbon can and does exist in the outer mantle. It can be brought up without becoming oxidized; it is associated with a variety of hydrocarbon molecules, both within inclusions in diamond and also in other materials brought up in the eruptions. Volatile-rich regions exist in the mantle, so that high pressure gas bubbles become assembled there that can force their way violently through all the overlying rocks. This clearly shows that the Earth has an unmixed, inhomogeneous mantle and that there is a high concentration of carbonaceous material in many areas of the globe.

Horizontal and Vertical Patterns of Hydrocarbon Fields

We have discussed the helium association of petroleum as an effect that points to a deep origin of petroleum. Everyone nowadays thinks of Arabia, the Persian Gulf, Iran, and Iraq as being the principal oil region of the world. It is indeed one connected large patch that is oil rich, stretching for 2,700 km from the mountains of Eastern Turkey down through the Tigris Valley of Iraq and through the Zagros Mountains of Iran into the Persian Gulf, into the plains of Saudi Arabia, and further south into the mountains of Oman. There is no feature that this large region has in common that would give a hint why it would all be oil and gas rich. The oil deposits are in different types of rock, in rocks of quite different ages, and they are overlaid by quite different caprocks. It cannot have been a matter of chance that this connected region had so prolific a supply of oil and gas, but in detail resulting from deposits derived in totally different circumstances in different parts of the region. These formations were laid down in such different times that there would have been no similarity in the climate or in the types of vegetation that existed then, and there is no similarity in the rocks that now contain the oil or in the caprocks that hold it down. Yet it is a striking fact that the detailed chemistry of these oils is similar over the whole of this large region. Surely, this is an example of the need to invoke a larger scale phenomenon for the cause of the oil supply than any scale we can see in the geology of the outer crust? Oil-rich regions seem to be defined by much larger scale patterns than those we see in the surface geology or topography or in the nature of the sediments of the region.

How Do Liquids or Gases Make Their Way Up Through the Crust?

Fluids that are liberated in the mantle and that are of lesser density than the rock will be buoyant relative to it, and if they are present in sufficiently large quantity, they will make their way up in one way or another. One way is in volcanic regions, where magma, with its density closely similar to that of the surrounding rock, holds open vertical pathways down to depths of a hundred kilometers or more. Any other fluids that are less dense than the lava and that have made their way through cracks into such lava channels will then move upward and quickly reach the surface.

Fluids can come up also in nonvolcanic regions. As buoyancy forces drive them upward, fractures will be opened or created, through which the fluids can move up.

It is clear: no rock can arrest the flow to the surface of a sufficiently large mass of a lower density fluid, and since the rocks have virtually no tensile strength, they cannot hold down any pressure greater than that given by their rock overburden weight.

If the fluids have originated at great depth and pressure, as I believe is the case with hydrocarbons, then a steady flow upward may be set up over long periods of time. In that case, a simple rule applies: the longtime average of this flow cannot be changed by any circumstance at shallower levels. If a good caprock exists there, it may have dammed up the flow, thereby creating a reservoir zone, but then, the pressure in that reservoir would rise as more fluid is supplied from the deep levels, and eventually, this pressure will crack the caprock and thereby allow the mean flow rate to become again equal to that emitted by the deep supply region. Whatever the permeability of the intervening caprock may be, the flow out of the surface will depend only on the initial supply rate. Measurements on or near the surface of the flow rates of hydrocarbons are therefore a meaningful exploration tool, and this justifies the numerous measurements that show such flow rates to be so high that the largest oil or gas field would be exhausted in less than a million years. It also means that oil and gas fields will replenish themselves as production diminishes the pressure in the uppermost domain. This replenishment will be not just by the mean flow rate (which would still be insignificant commercially, taking perhaps 10,000 years to refill a reservoir), but a much faster refilling rate will result from the pressure change that production has caused. The pressure in the uppermost domain having decreased, the pressure in the domain below will now become greater than the rock can bear, and the top domain will get access to the next one below. This kind of refill will be one of great commercial importance, and several reports exist already of observation of this effect.

I published many articles in scientific journals giving these arguments for a nonbiological origin of petroleum, but it was clear that this viewpoint was not going to be accepted. These papers were well received by some of my peers, but were regarded as crackpot stuff by many others. One of my early supporters in the gas and oil industry was Robert A. Hefner III, a man who had been chiefly responsible for the discovery of large amounts of natural gas at great depth in Oklahoma. He had noted an article I had published in the Wall Street Journal in 1977, on the possible origin from depth, of oil and gas on the Gulf of Mexico coast. He wrote me a letter to say that my views would better account for his findings than the conventional ones. Could I meet with him in the near future? I did, and he has become a good friend and a close collaborator. But even with his great commercial success behind him, and his training as a petroleum geologist, he could not change the scene for me.

Two other important supporters were introduced to me in a remarkable way. While I was working for a semester at the Niels Bohr Institute in Copenhagen in the early 1980s, I had a phone call from Vienna, from a member of the International Institute for Advanced Systems Analysis (IIASA). The caller's name was Cesare Marchetti, and his message was the following: "In the predictions I have made for the availability of different fuels in the future, it has become clear to me that natural gas will be the next major energy fuel, surpassing coal and oil. I am glad that you have given the reasons why this will be so." He then told me that a friend of his, Ed Schmidt, would shortly be in Copenhagen, and he suggested I meet with him.

A few days later, Ed Schmidt turned up, and I had a few hours of discussion with him. He was a remarkable man; he knew and understood a lot in many fields; he spoke with great precision to the degree that one might have considered this an affectation (but I did not). He had retired and was now just following his interests. Among those, my view on the origin of petroleum and natural gas had now become a major item. He urged me to meet with him again and with Marchetti in Vienna. Marchetti turned out to be also a delightful person, widely read and knowledgeable in a wide range of the physical sciences. His job was to deduce the future from a knowledge of the past. That, of course, is the principal function of the brain for anyone, but he had made a special study of how to proceed for best results. It was in this line of work that he had realized that natural gas would soon match and then surpass oil as a fuel, and from the rate of rise of gas usage, he had concluded that there must be a lot more around than people had realized. My viewpoint then gave him greater confidence that this was indeed so.

I acquired several other supporters, both in the petroleum industry and outside of it, but especially in the Soviet Union where a lot of work had been done along similar lines. On one of my numerous trips to the Soviet Union at the invitation of the Soviet Academy of Science, I was asked to give a colloquium on this subject in Moscow. I had a large audience that included several senior members of the Academy. In contrast to lectures I had given in the West, there were no hecklers but much positive comment. Zel'dovich, the famous physicist, was there, sitting next to Levin, a scientist whom I knew well, and who told me later some of the comments Zel'dovich had made to him. The most interesting one was: "Professor Gold must be in a very powerful position, that he can speak so strongly against the established opinion in his country." Unfortunately, he was not right about this; I was just willing to accept the risks that my outspokenness entailed.

The Search for Hydrocarbons in the Bedrock of Sweden

I had by now reached a stage where I felt what I had to do was to make a clear, practical demonstration of oil or gas production on a commercial scale, in a location where the biotheory could not possibly have provided an explanation for the existence of these fluids. Nothing short of that would turn heads. I thought of this as a subject of enormous economic importance, firstly because the prognosis for future supplies had great influence on the global oil–gas economics; secondly because new thinking and new exploration techniques would be indicated, through which many new discoveries might be made all around the globe. My message was: "Do not think that just the areas now known to be oil and gas-rich are the only ones. This precious commodity will be seen to be far more widespread around the globe than was previously considered, and the dependency of many countries on oil importation might be greatly reduced."

How then could I go about finding hydrocarbons in commercial quantities in new places, places that were quite unexpected in the conventional theory? An opportunity presented itself unexpectedly. Mr. David Bardin, a lawyer in Washington whom I knew and who had served under President Carter as a senior administrator in the Department of Energy, had developed an interest in my theories and made the contact with persons in Sweden that resulted eventually in an invitation for me to present my theories to a group at the Swedish State Power Board. In the mid-1970s, when I had gone more deeply into the subject of the origin of hydrocarbons, I considered Sweden a particularly good location in which to test my views. I made my presentation on June 2, 1982. At first, I sensed some hostility to my views, but then suddenly, the atmosphere changed, as a result of one participant displaying great interest and adding information favorable to my case. I discussed not only the general viewpoint I had reached about the origin of hydrocarbons on Earth, but also the particular interest this would have for Sweden, where the conventional viewpoint would virtually deny any possibility of finding hydrocarbons, but where, on the contrary, on my basis, the situation was particularly favorable. I pointed out, as a prime area to investigate, the circular feature called the Siljan Ring, a very large meteoritic impact crater; it is in Central Sweden, seen on the map as a ring of lakes enclosing a circular area of approximately 35 km diameter. The ring itself is a depression, a few kilometers wide that contains a shallow deposit of sediments, nowhere deeper than 300 m.

I gave my reasons for regarding this structure as being of prime interest: an impact that made a 45 km crater would have smashed the entire thickness of the crust, thereby generating pathways from great depth and leaving porosity in the crushed rock that had welled up after the impact and had filled the crater.

The plan I proposed was to perform first a geochemical study of the entire circle. This investigation involves checking the area for the various chemical and isotopic signs that are known wherever a long-term substantial seepage of methane and other hydrocarbons has taken place. A positive result there would be an extremely strong indication and would make the case for some seismic profiling and exploratory drilling.

Following this, I was invited to Stockholm for a further meeting. It was agreed that a few samples from the Siljan region would be collected before the snow and that they would be sent to me to arrange for chemical analysis. I asked for soil samples from a depth of the order of 1 m, from locations distributed over the region interior to the Siljan Ring, and for some samples of exposed rocks from the Ring. The soil samples turned up calcium carbonates (limestone), of a type characteristic of a hydrocarbon origin of the carbon, unlike the common marine limestone. This demonstrated that the seepage of methane was still active in Siljan. The carbonates from the ring sediments showed the ordinary composition of marine sediments, in accord with the marine fossils they contained. The oil from the oil seeps had a composition that was quite typical of a biodegraded ordinary crude oil.

The highlight of the day of exploration was a look at the oil seeps and the oil standing in drill holes at one of the two limestone quarries. However, at that time, we did not know that there was much more oil standing in several other drill holes at the quarry and also oozing up in several locations on the level working ground, as well as dripping gently from cracks in steep walls. The two known areas of oil seeps were both in the sedimentary ring, and they were only in the two limestone quarries, about 10 km apart. In the interior of the ring, there was no limestone and therefore no quarries. The oil had been regarded by many geologists as arising from organic sediments in the ring, while others had objected that so small a sedimentary deposit could not be held responsible. Similar seeps in settings not as clean as walls of white limestone, but just in the forests of the region, would not have been discovered. I could not suppose that the seeping oil had just found its way into two quarries where it would show and not into the ground between them where it would not have been observed before being bacterially decomposed. The quarries and their boreholes were not deep, and so they could not have constituted particularly favorable spots for seepage, but only favorable ones for their detection. Even if the oil had just come from the ring sediments as some would have, one would have to suppose that there were many other seeps unnoticed in the forests between them. Any estimate of the rate of total seepage, even just in the ring area, was large enough to exhaust the largest oil fields of the world in a few tens of thousands of years, but the impact feature itself was dated as 360 million years old, and it would seem unlikely that it started to leak out oil just in the last 10,000th part of its age. But if the oil did not originate only in the ring, but overall the area of the impact feature, there would have been many times as much. On the occasion of a later visit with some of my friends from Oklahoma, I overheard the remark: “Gee, if we had seen seeps like these in Oklahoma, we sure would have drilled the hell out of them.”

Siljan now seemed an extremely good prospect to me, but I was told that we simply had to wait for the approval of our report before the next round of investigations could begin. At the meetings, there was good participation by the Swedish Geological Survey, and we were informed that the director of the survey was favorably inclined to the project, despite its novelty, and so were the academic geologists in Uppsala University who had previously made a study of the Siljan area and had established its origin as the site of an impact. On the other hand, it became clear that the Swedish Petroleum Company had declined any participation, and one could safely assume that in the higher circles of the Swedish government, they would express opposition to the initiative.

A series of surface chemical investigations were undertaken along the lines that I had proposed. These were to find out whether indeed the Siljan Ring area was rich in hydrocarbons, and if those turned out positive, then there would be a follow-up by exploratory drilling and then serious large-scale drilling. At my suggestion, the Swedes invited a US chemical exploration company, Geochemical Surveys, Inc. of Dallas and Denver, to do an extensive soil sample survey for hydrocarbons and certain elements known to be associated with hydrocarbons. This again turned out very positive. It demonstrated clearly that the various residues of oil and gas seepage were highly concentrated within the Siljan Ring impact structure, and the levels of the various substances considered as petroleum indicators were as high as in any petroleum-bearing area in the USA. Anybody looking at the data would have no doubt that, indeed, just the impact areas, up to and including the ring with its depression and lakes, were just the favored regions for gas and liquid petroleum seepage. My feeling was, at that stage, that I could not possibly be wrong. To have put my finger at this spot on the map and then to find that it clearly was an outstanding high spot in hydrocarbon seepage would be too much of a coincidence if it were based on the wrong theory. However, others did not see it in that light. Some told me that all the hydrocarbons and their residue products could be produced from the top soil which, at present, is only about three meters in most of the area, but which, of course, may have been deeper in preglacial times. Why this would accentuate the impact area was not discussed by these critics, nor why there should be liquid oil seeps in an area of extremely small (less than 300 m deep) sedimentary deposits was also not made clear. "It is easier to explain it as a result of present and previous sediments than to explain all this on the basis of Professor Gold's farfetched theories." That was the kind of response that I heard quite frequently. Easier to explain of course just refers to the framework within which this explanation is placed.

But I had to do the best I could to get a deep hole drilled in this most attractive location. I would much have preferred to have the whole project run by a team that included several persons from the USA. The Swedish response to this was "It is true that we do not know much about drilling for oil and gas, but we will learn fast," a response which I found absolutely appalling. Here, I was offering them a cooperation with persons who had discovered many billions of dollars worth of gas in the USA, and they refused this on the basis that they could learn for themselves. Knowing how long it had taken to learn all of the numerous difficulties encountered in deep drilling, I considered that an excessively brash remark. Indeed, that is what it turned out to have been.

In 1985, five exploratory holes were drilled by the method of continuous coring and down to a depth of approximately 500 m. These holes were distributed over the interior of the Siljan Ring, one of them being in the shallow sediments on the eastern edge. All five holes immediately demonstrated the presence of hydrogen, methane, and heavier hydrocarbon gases and also the presence of the carbonate crack-filling cements with which I had become familiar in other hydrocarbon seepage areas.

This carbonate result was important not only for demonstrating a similarity to other known hydrocarbon areas but also for giving an estimate of the quantity of hydrocarbons that must have gone through the ground. We observed a concentration of this particular type of carbonate cement down to at least a depth of 500 m and apparently distributed over the whole area of the interior of the Siljan Ring. One could therefore calculate how much methane must have become fixed during its movement through the upper granite. Presumably a much larger quantity had in fact escaped. But in any case, this quantity by itself turned out to be enormous: it was obviously in quantities far outside anything that could be attributed to biological debris laid down in the area.

Despite a great deal of skepticism and prevarication by committees, a drilling program commenced. The funds for drilling came in major part from the Vattenfall energy company in Sweden, plus significant contributions also came from Swedish investors and from the American Gas Research Institute, which was interested not so much in the commercial success of the enterprise but in the scientific findings that would come out of this very unconventional drilling. It was proposed to drill to a depth of 7.2 km at which a discontinuity in the seismic profile was seen that could be interpreted as the upper boundary of a high pressure oil field zone. The rig was one of the largest and would in principle be capable of drilling to such depth. The site chosen was in the interior of the ring in the north-northeastern sector, chosen because this was both favorable according to the seismic information, but also the gravity information indicated that the underlying rock was of lower density there, which could be interpreted as possessing more porosity and therefore be more favorable for the extraction of oil or gas. Drilling was started at the beginning of June 1986.

In June 1987, just one year after the beginning of the operation, a most remarkable sequence of events occurred. As a result of a drilling mishap, the drill was stuck at a depth of six kilometers for a period of 10 days. When it was finally freed and brought up to the surface, the lowest 10 m of the drill pipe were tightly blocked by a very stiff paste. The material was black, about the consistency of putty, and it had a strong and objectionable smell.

As soon as I received information about this find, I asked for a sample of the material. There was no reply and no sample came. I wrote similarly to the on-site chemist also with no success. The Vattenfall people had in fact sent a sample for analysis to the Norwegian Petrochemical Laboratory, Geolab Nor, and the reply came back from there that showed the oil of the black paste to be chiefly a light oil with smaller amounts of heavy molecules whose precise identity was not established. Geolab Nor stated that this oil did not show any resemblance to any of the drilling additives that had been supplied for comparison. Three months elapsed after the discovery of the material before I was able to obtain a sample. Then on a visit to the site, I insisted to receive a sample before I left. It turned out that although some 60 kg of this thick, black paste had been bored out of the pipes when they were finally brought to the top, almost all of it had been thrown away. All that remained was one small plastic bag with this material in it. The on-site chemist told me that the smell indicated that it was some bacterial product, and therefore, it must be something that had fallen in from the top of the borehole. How 60 kg of a uniform black paste would have fallen in at the top was not described, nor why it would have oil as a binder, rather than water, which was the drilling fluid at the time. All but the small sample in his hand had been bulldozed into some ditch and covered with dirt to get rid of the smell. The small sample he had retained was in an ordinary domestic polyethylene bag, and polyethylene soaks up and transmits oils. A chemist might have known this. The result of this inadequate preservation was that the material was now rather stiff and not as pliable as it had been described initially.

It so happened that I was to go directly from Sweden to a brief vacation on the Spanish Mediterranean island of Mallorca. I arrived there on a weekend, and for one and a half days, I had no access to anything one might buy in a hardware or a drug store. Yet I was fascinated by the black material and could not wait that long before analyzing it. I looked through the apartment for anything that might serve as an oil solvent, but there was no paint solvent, no nail polish remover, or anything of that kind. There were magnets in the house, namely, the magnetic door latches of cabinet doors. I unscrewed one and ascertained that the mystery material was strongly magnetic. Then I put a small amount of the material in hot water and kitchen detergent, and indeed, after some effort, it dissolved. The dilute liquid so produced was almost transparent, just had a slight gray, hazy appearance. It evidently contained particles, but they were very small. I then put a drop of this fluid on a piece of aluminum foil and it looked completely transparent. Then I held the door latch magnet underneath, and instantly, the two lines of the magnetic poles appeared as completely black lines on top of the foil. Similarly, when I held the magnet to a side of a glass filled with this liquid, it produced immediately a large black patch on the wall of the glass. Further experimentation demonstrated that essentially this material consisted of little other than very small magnetic grains and an oil that could be dissolved in water with kitchen detergent.

So what was all this? The only black and magnetic material of which I knew that occurred in nature was magnetite [Fe_3O_4]. So what I seemed to have here was quite unusually fine-grained magnetite, yet in the size range large enough to be ferromagnetic. The strong smell was a mixture of the smell of a heavy oil together with rotting biological material. A dead rat on the garage floor would be the best description I could give! I judged that a crude oil had been available at six kilometer depths in the granite and had somehow concentrated the small magnetic grains. Microbial action was indicated by the small size of the grains and by the strong odor. To be sure about the identification of magnetite, I sent a sample to a friend in Copenhagen, Dr. Jens Martin Knudsen at the Ørsted Institute of the University of Copenhagen, for analysis by Mössbauer spectroscopy. Within 2 days, he had established that the material was indeed fine-grained magnetite. He also said that the spectrum showed a deformation, which has to be identified as due to the presence of another metal within the crystal lattice, most probably zinc, at a concentration of about two percent. The presence of zinc within the crystal lattice was an important clue, since there was no zinc in any of the drilling hardware or in any of the additives. So the zinc must have been available in the formations of these crystals. I remembered that in the vicinity of the Siljan Ring, there had been a commercial zinc and lead mine.

There ensued a long debate about the origin of the magnetite. Not only the chemists but several other persons associated with the project considered that it must have fallen down, having somehow gotten into the hole on top. It seemed to them just impossible that nature would have produced such a substance at depth. Entering from below, the black paste must have been at a higher pressure than the drilling fluid, for it had gone a long way up in the drill pipe.

A large international conference had been organized to take place in the vicinity of the Siljan drilling, and delegates from all over the world were present, including a large contingent from the USA who were engaged in the preparation of US deep drilling for scientific purposes. The Swedish holes were by then the deepest outside of the Soviet Union drilled in nonsedimentary rock. The magnetite mystery clearly should have represented a major source of information for the conference, and the observations of hydrogen, helium, and hydrocarbon gases should have been presented as a major item of interest. Instead, the whole discussion seemed to center around the failure of the enterprise to produce commercial quantities of gas. Drilling had been stopped for reasons of shortage of funds, just a few days before the conference, and this led to a triumph of the antagonists of the project. People came away from the conference with the impression that they had heard a story of an expensive, foolish, ill-advised attempt, organized by persons who had no acquaintance with deep drilling or petroleum, and upon the advice of a person who had no standing in the field of petroleum geology [myself].

Drilling resumed after a short interval, and a total vertical depth of 6,764 m was reached. But then the borehole became too unsteady, drilling had to stop, and one had to case the borehole. When by April 1989, it had become clear that the drill could not reach a greater depth than 6,700 m. I expected that at least there would be the long delayed detailed testing at various levels for oil and gas that might enter the well bore. It would be a very important occasion, and I was confident that the result would be positive, since we had already seen a crude oil in the hole.

Matters then took a turn with which I thoroughly disagreed. In order to extend the borehole by hydrofracking, a heavy fluid was poured into the hole. This achieved nothing other than to obscure completely the lowest kilometer. So we now had effectively a hole to a depth of about 5.5 km instead of 6.7 km. What could we do with this hole now? I proposed renting a down-hole pump to see what fluids we could pump up from the accessible levels. This was done, the pump was installed, and it was used for about a month. It pumped up first many tons of saline water, but then, it changed over to pumping up oil and simultaneously with that small grains of magnetite like those we had seen before. Immediately, an explanation was offered for the presence of the oil by several people who were informed of all that had gone on at the site, and the suggestion was made that the oil that we saw was not a crude oil but instead was a diesel oil that had been a component of an emulsion used as drilling fluid. I disagreed. I obtained a sample of this oil and had it analyzed in some detail and compared with the diesel oil that had been used in drilling. There were very substantial differences, and on this basis alone, one could rule out that we were looking at diesel oil.

A further investigation made the case even clearer. The recovered oil contained significant amounts of nickel and vanadium, the two elements that commonly occur in natural petroleum but that would be absent in any distillation product like diesel oil. Also, the oil had some prominent heavy components which could not possibly be tolerated in a diesel engine. Nevertheless, once the word had gone out to the press that we were all fools and that we were just studying oil that we had poured into the well, it was almost impossible to get any journals or the press to listen to these details. All our antagonists went out of their way to ridicule us for not realizing what the simple explanation was and for trying to persuade the world that a discovery had been made. No journal would publish these results, and the newspapers and news organizations gave it a brief mention, mainly to ridicule our efforts. The two deep-drilling operations outside the USSR were in Germany and the USA, and I approached both, asking whether they would like to send a representative to Siljan to look over the operations and perhaps to take samples. The hole was by far the deepest drilled in solid bedrock, far deeper than anything available at that stage to these two organizations. I thought that they would be absolutely obliged to see first what the results had been in the deepest hole in granite, before drilling very expensive holes themselves. The answers that came back from the Germans were that they were too busy and could not spare any time, and from the American operation, that they had no objection to my listening in at the next conference they were planning, provided I paid the [substantial] registration fee.

The pumping operation was expensive, and although 12 tons of oil were pumped up, it certainly did not look as if a commercial production from these levels could be obtained. It seemed to me that the operation had been completely successful in demonstrating that crude oil exists at great depth in bedrock and far removed from any biological sediment. If it had come into the well-bore at a rate sufficient for commercial production, then I could have laughed at all the absurd criticisms. But one cannot sell 80 barrels of oil that it took 2 days to pump up, and therefore, as a commercial endeavor, this hole was not a success. The whole operation was generally described as a failure, and the fact that 12 tons of crude oil had been obtained from great depth in the granite went unnoticed. It went unnoticed even by the people who had previously stated that it was absurd to drill there since "there could not possibly be any oil down there." If we had seen a single drop of oil down there, it would have confirmed that oil can exist at depths in granitic rock. Twelve tons at an arbitrary location in the impact crater seemed to me to make a very clear case.

The story of the black magnetite grains continued. About 14 tons of those were obtained in the pumping operation together with the oil, and it slowly became clear that those grains were the chief culprits in preventing a better flow of oil into the well-bore. The small cracks through which fluids normally enter the bore seemed to get choked up and concentrate the black paste until nothing more could get through. It seemed to me an ironic turn of fate that the magnetite paste, which I regarded as a confirmation that bacterial activity had taken place down there, became the obstruction to commercial exploitation.

We still had to show that the content of biological molecules present in the oil was really due to microbiology flourishing at depth. The very-fine-grained magnetite was presumably a product of microbial action. The smell which our chemists had identified correctly as resulting from decaying organic matter seemed to me to be a good indication of microbial action, but having taken place down hole. I therefore asked for the Swedish National Bacteriological Laboratory in Stockholm to be invited to take samples from different levels in the well and to see whether they could culture some microorganisms from those samples. The laboratory microbiologist cultured three different strains of bacteria, strains that had not previously been seen but that were clearly related to others that were known. These bacteria reproduced only at elevated temperatures [thermophilic], and they could not tolerate free oxygen [anaerobic]. These circumstances corresponded closely to those at the levels from which they had been sampled, and there were no other bacteria that they could culture from these levels. Contamination from the surface therefore seemed very improbable.

I did not think that this procedure used would allow the primary organisms to be cultured, those that produced the magnetite by feeding on the combination of hydrocarbons and more highly oxidized iron. From the composition of the oil that was pulled up, one would conclude that the bacterial action had removed chiefly the lightest hydrocarbon molecules, such as methane, and since these are very compressible gases, their density down hole at the high pressures would be hundreds of times greater than at the surface. A high pressure transfer and culturing apparatus would be required to culture them, but this was not available. The bacteria that were cultured presumably belonged to a second step in the food chain.

The Second Hole

Despite the lack of a commercial success, there were enough Swedish investors who had understood the significance of the result and also the absurdity of some of the actions that had been taken. They formed a company and obtained the mineral rights from the Vattenfall energy company. This new company made quite clear that they wanted to be guided by experienced US drillers. The decision was taken to drill exactly in the center of the Siljan Ring in a location called Stenberg. In this hole, lubricants based on natural petroleum were totally prohibited. Great care was taken with the analysis of any substances that were poured into the hole. A company in Oklahoma, with which I had worked before, was appointed to examine the drill cuttings from every 5-ft interval of drilling. The examination involved extracting the hydrocarbons that may be in the interior of each grain, as they would be if the pore spaces had been loaded with hydrocarbons for a long time. This meant that hundreds of such values had to be measured, and many persons thought that this represented a great waste of money and time. However, as it turned out, the results were of considerable importance as they left little doubt that the hydrocarbon fluids had all come from below any of our drilling depths.

One day, I received the exciting news that very high values of hydrocarbon content were observed at several depth intervals in the second hole, intervals that represented in each case the penetration through igneous intrusive dolerite rock. Instead of the light-colored granite chips, there was a mixture of those with black chips or sometimes only black chips. The hydrocarbon values in these intervals were as high as any previously seen in the best petroleum-bearing areas. The interpretation of this close association with the volcanic intrusive rock clearly implied that the hydrocarbons had come by the same pathways as those by which the intrusive rock had come earlier. There is no question that the intrusive rock had come from below and not from above, so this alone meant that we were indeed seeing hydrocarbons coming from greater depths. The reason why the intrusive rock might provide better pathways from deep levels is twofold. Firstly, the intrusive rock coming from below will have intruded in the granite in the location where it was weakest. Secondly, the intruding rock which would freeze in the colder granite would then be locked to the surrounding rock, but still on further cooling, it would contract. This contraction would tear up and make cracks both in the intrusive mass and in the granite closely surrounding it. That would then account for the presence of the hydrocarbons in each of five cases being in the dolerite but also in the granite just adjacent to this dolerite.

These were very exciting times, holding out the possibility of commercial production of hydrocarbons, but showing in any case that a lot of hydrocarbon fluids existed in the Siljan crater. If one took the results of the first and second hole, recognizing that neither was drilled in any location that was more specifically identified than the average of the crater, I think one would be entitled to suppose that those measurements would have applied to the entire crater. In that case, just taking the hydrocarbon content that had been measured in those holes and multiplying it on the basis that such layers would on average be the same proportion of the rock volume as they had been in our two holes, one came to a figure which by orders of magnitude corresponded to some of the largest oil fields in the world.

Our aim had been to drill down to 7.2 km depth on the basis of our seismic information. Unfortunately, the same drilling problems were encountered as those that had stopped further drilling at Gravberg. When the drill had reached 6.5 km depth, drilling had to be stopped. The friction of the drill string against the rock with its numerous breakouts had slowly come close to the value at which it would break. The last thing one wants is to have a great length of broken steel pipe standing in the hole, which it would be very difficult or impossible to remove. Drilling was stopped, and the instrumentation was put in place to test the outflow that could be obtained from the individual high spots of hydrocarbons that had been observed. Such testing requires that a casing pipe is perforated in the location to be tested and then that it is plugged in such a way that the total pressure can be greatly reduced below that of the drilling fluid in the pipe. It is only then that one can judge what rate of inflow can be obtained. This was the kind of test which had not been done in the first hole, and now we had arrived at this crucial point. The test results showed an inflow of both oil and gas in the selected locations. The only trouble was that the inflow rate would rapidly decline to a low value. The pore spaces that led to the inside of the well bore were evidently getting blocked quickly by some minerals carried in the oil. We later found out that this was the same fine-grained magnetite as we had seen in such large quantities in the first hole. The excitement was over. We could still not produce commercial quantities, but we had oil, a canister of free gas, and again a large quantity of the magnetite. The oil was analyzed by several petrochemical laboratories, including those of the Danish Geological Survey, and they all told us that this oil was a perfectly ordinary crude oil that had suffered some biodegradation. We were now seeing just the same oil with the same distribution of the hydrocarbon molecules as we had seen in the first hole. The criticism we had encountered—that the oil found in the first hole was due to diesel oil that had been used 4 months earlier, and that had somehow modified its composition—was wrong. This hole was 12 km from the inner edge of the ring and surely nobody can seriously suggest that a 300-m sediment in the ring would cause oil to flow into the granite to a depth of 6 km and a horizontal distance of 12 km.

Although the second hole had completely confirmed the deductions we had made from the first, all this information could not be published and could not even be brought to the attention of the people who were conducting the large drilling operations in the USA and in Germany. Although it was clearly a remarkable result to find natural gas and oil at depths of six kilometers in bedrock and to find bacterial life and its residue products there, we could not get this published in any of the major scientific journals, nor did the popular scientific press pick it up. The reason was that the opposition to a nonbiological origin of petroleum was fierce, and the concerted opinion was that our results "were just not possible and therefore had to be discarded." It was a very distressing situation for me to have found so clearly the deep origin and the microbial contamination of oil at depth in the rock, and yet be unable to inform the geological community of this.

What did the Swedes get out of all this? They obtained the knowledge that their country could become a major producer of oil and gas, but only after the extreme prejudice against a nonbiological source of petroleum had been brushed aside, and new efforts of drilling in bedrock could be started again. The difficulties we had in trying to obtain commercial flow rate were not of basic and general importance, and work based on our experiences would overcome the problem. The oil and gas are there in abundant quantities and will one day be produced. The problems were not in the ground; they were in men's minds.

Chapter 11
Earthquakes

The natural sudden events that are most devastating for humanity are the earthquakes. Thousands of persons have been killed in seconds or minutes in the violent events that our planet suffers. What causes these events and are there any ways of predicting them? Many lives could be saved if a good prediction system could be found. That is what I want to explore in this chapter. The most common theory of earthquakes, and the one to which the US Geological Survey subscribes, is that subterranean forces of unknown origin gradually build up strains in the crustal rocks, up to their breaking point. When that point is reached, the rock breaks, and it is the sudden brittle fracture that causes the shocks of the quake.

Many accounts of events of past earthquakes tell quite a different story. Frequently, we find references to massive emissions of gases from the ground, both before, during, and after a quake. Are such gas emissions merely a consequence of the fracture of the rock? How could they ever precede the quake? Why would there ever be many earthquakes in succession? These and other inconsistencies of the strain buildup theories with reports of many past earthquakes had caused my colleague Steven Soter and me to become very critical of this favored viewpoint.

But in modern times, not much attention is paid to accounts of past quakes, and not much discussion of them is in current seismological literature. Such information is considered as hearsay and uncertain. Instead, the effects that can be measured with accuracy, such as a gradual increase of the strains in the rocks, and their relation to earthquakes have now become the main subject of research in this field in the USA. The overriding importance of this research lies in finding a method for the prediction of earthquakes, but so far, strain measurements have not been found to have any predictive value.

In the past, there has been one outstanding event where prediction was very successful. On February 4, 1975, the city of Haicheng in Northeast China was evacuated just 2 h before a massive earthquake struck. But this prediction was not based on accurate strain measurements in the rocks or anything of that kind. It was based entirely on gas-related phenomena. For weeks prior to this earthquake, the air temperature in the vicinity of the local fault line was higher than in the surrounding

Thomas Gold *edited by* Simon Mitton, *Taking the Back off the Watch*,
Astrophysics and Space Science Library 381, DOI 10.1007/978-3-642-27588-3_11,

region, and this difference increased at an accelerated rate up to the day before the earthquake, when it reached a temperature differential of as much as 10°C.

We read the following report from the Liao-ling Province Meteorological Station (1977), responsible for the Haicheng area:

During the month before the quake a gas with an extraordinary smell appeared in the areas including Tantung and Liao-yang. This was termed "earth gas" by the people . . . one person fainted because of this. . . . Many areas were covered with a peculiar fog (termed "earth gas fog" by the people) just prior to the quake. The height of the fog was only 2–3 m. It was very dense, of white and black color, non-uniform, stratified and also had a peculiar smell. It started to appear 1–2 h before the quake and it was so dense that the stars were obscured by it. It dissipated rapidly after the quake. The area where this "earth gas fog" appeared was related to the fault area responsible for the earthquake.

My colleague Steven Soter and I found the Haicheng explanation convincing for our viewpoint that gas emissions coming from below were a major component of the earthquake phenomenon. The triggering of earthquakes was not due to the strain having grown to breaking point of rock, but rather that gases had entered and greatly weakened the strength of the rock. These same gases may previously have mixed with and driven out of the overlying soil gases normally present in the porosity at such levels. The gases below a few meters depth would have been about ten degrees warmer than the surface air in midwinter, and saturated with water vapor. They therefore produced a fog on contact with surface air. Despite its warmth, the gas mixture coming out of the ground would remain close to the ground if it were more enriched in CO_2. An enrichment of about seven percent would suffice to cause people to feel unwell, and emission of gases may also have accounted for those several incidences of anomalous animal behavior and changes in groundwater reported prior to the earthquake itself.

After learning about the Haicheng earthquake and its precursor events, Steven Soter went to work to find a large number of historical events, many of which seemed to require similar explanations. The most dramatic of these accounts comes from the town of New Madrid, on the Mississippi, where a series of earthquakes struck on December 16, 1811, followed by one on January 23, 1812, and by yet another on February 7, 1812.

One contemporary report by W. L. Pierce said:

During the first four shocks, tremendous and uninterrupted explosions, resembling a discharge of artillery, were heard from the opposite shore [of the Mississippi River]. . . . Wherever the veins [fissures] of the earthquake ran, there was a volcanic discharge of combustible matter to a great height, an incessant rumbling was heard below, and the bed of the river was excessively agitated, whilst the water assumed a turbid and boiling appearance. Near our boat a spout of confined air, breaking its way through the waters, burst forth, and with a loud report discharged mud, sticks, etc., from the river's bed, at least 30 ft above the surface.

There were many eyewitness accounts of the 1811–1812 New Madrid earthquakes, and the curious reader can readily find these on websites. I will briefly note some of the phenomena connected with these large earthquakes, the state of the weather at the time, and the opinions which the people held in regard to their origin, or the great cause of them.

The weather was warm and smoky, and had been so for some days, not a breath of air stirring, and so thick and smoky that the Kentucky shore, one mile distant, could not be seen at all. They were in a balmy Indian summer. The morning after the first shock, as some men were crossing the Mississippi, they saw a black substance floating on the river, in strips four or five rods in breadth by 12 or 14 rods in length, resembling soot from some immense chimney or the cinders from some gigantic stovepipe. It was so thick that the water could not be seen under it. On the Kentucky side of the river, there empties into the Mississippi River two small streams, one called the Obine, the other the Forked Deer. Lieutenant Robinson, a recruiting officer in the United States army, visited that part of Kentucky lying between those two rivers in 1812 and states that he found numberless little mounds thrown up in the earth, and where a stick or a broken limb of a tree lay across these mounds, they were all burnt in two pieces, which went to prove to the people that these commotions were caused by some internal action of fire.

In the Mississippi River, about five miles above what was then called the first Chickasaw Bluffs, but in later times Plum Point, was an island about three miles long, covered with a heavy growth of timber, which sank in one of these shocks to the tops of the trees, which made the navigation extremely dangerous in a low stage of the river.

The Saint Francis Swamps seem to play a major role in the New Madrid earthquakes; several contemporary reports refer to them. The most common explanation for the presence of swamps is that they represent an area of no drainage, and therefore, the water that accumulates in them will eventually be loaded with vegetation and bacterial debris whose decay will use up the oxygen in the water and therefore make it anoxic. In anoxic water, any growth that falls in will no longer be subject to the decay mechanisms that are present in oxygenated water where carbon in plant debris is then turned essentially into carbon dioxide which, of course, escapes from the surface. Some bacterially produced methane may also be present from anoxic vegetation decay. The remaining carbon fibers are the most resistant of any of the chemicals of the plants, and therefore, the water will become more and more filled with them, thereby damming up any outflow and supporting only specific swamp vegetation.

However, I have seen swamps which were not set up in this way. They were in places where there was no drainage difficulty so they could not in the first place become anoxic, because there would not be any long-term stagnant water. There are swamps on steep hillsides where the water could run off freely, but nevertheless, some particular area grew into a swamp land and then did hold up water. Another mechanism must have started the anoxia of the water which will then perpetuate itself as new incompletely decayed plant material becomes available in them. What was it that caused the anoxia in the first place? The clue we found was that these swamps overlay a fault line of the underlying rock quite precisely, and it is therefore a reasonable assumption that hydrocarbon gases have entered the water and caused the anoxia. Bacteria that need oxygen for oxidizing methane will multiply much faster than those that live on oxidizing plant debris, and therefore, stagnation of the water is maintained.

Of course, since methane will also be produced in such a swamp, the finding of methane gas is usually not thought to be significant. By modern techniques, however, one could easily check which of the mechanisms was at work, because natural gas, although mainly composed of methane, almost invariably has also the higher carbon number gases: methane, butane, propane, as well, while bacterially produced methane is just plain methane.

The Saint Francis Swamps may have arisen in the second way, and this could probably still now be tested. It is usual for methane emission from the bedrock to continue still long after any period of massive emission, and one could check whether the other hydrocarbons are also present. That will then also explain that the reported sudden sinkage of a large area of ground was indeed due to gases having escaped from all the pore spaces in the rocks which they previously had held open, and that this caused all the chasms that were reported there.

The San Francisco Earthquake

The earthquake that destroyed parts of San Francisco and virtually all of Santa Rosa occurred at 5:12 a.m. on April 18, 1906. It was most intense, perhaps a 100 km north of San Francisco. We will here list some excerpts from the numerous reports, all indicating violent gas emission from the ground, gases that contained the poisonous hydrogen sulfide and gases that were frequently flammable. It is the earthquake for which the most detailed reports exist, and which shows every type of phenomenon that we have noted in other cases. The reader may like to examine the richness and variety of these reports by using a search engine. One important source is A. C. Lawson et al., *The California Earthquake of April 18, 1908.*

Lawson includes the following:

From Santa Rosa, "Heard noises in SW; then felt breeze; then felt shock." From Cotati, "Sound as of a strong wind before the shock." From Point Reyes Station, "Heard roar, then felt wind on my face." From Calistoga, "A rushing noise before the shock came." From Pescadero, "Noise as of wind preceded the shock." And from Mount Hamilton, "Sound as of flight of birds simultaneously with shock."

Other clear evidence for gas is given by a report published on April 23 in the Santa Rosa *Democrat-Republican* (the first newspaper to appear after the devastation). It said:

J.B. Doda, who came over from Fort Ross on Monday, reports that the earthquake caused immense cracks in the earth there, from which strong gases are emitted which make men and cattle sick.

Also, according to Edgar Larkin (1906), who collected a great many accounts, the odor of hydrogen sulfide was noted in the area of Sausalito. He also reported that sulfurous odors were pungent in Napa County during the night of the 17th and 18th before the upheaval, and lasted all day. ... From many of the letters it is clear that the entire region north and east of San Francisco is saturated with gases of sulfur origin. . . .

The phenomenon of slowly rolling waves, like the waves at sea, was reported from many places in the San Francisco earthquake. Lawson and others (1908) list over 20 such accounts distributed geographically from the vicinity of Eureka to Visalia, a distance of more than 600 km. Several of these accounts explicitly compare the ground motion observed to that of waves in the ocean. What I presume is happening is that soft alluvial deposits can be bent just like a floor carpet and do not fracture as readily as the hard rock beneath. If a great mass of gas suddenly comes from cracks in the rock, it may lift up this carpet, and in that case, gravity waves quite similar to the waves in the ocean would be set up.

Other Historical Earthquakes

Soter found many reports of major earthquakes in historical times, some of which will be appended here. Many, or indeed most of them, refer to the emission of combustible gases as a major feature.

One has to take these reports seriously, because many observers give very similar accounts coming from different parts of the world, and with no possibility of any communication between the different reporters. The gas effects that are observed very clearly in many cases seem to me to have a better chance of leading to a predictive capability than the strain measurements. After reading these and many other reports about, I find it very hard to understand how there can be any opposition to the conclusion that the eruption of gases is connected with earthquakes, and possibly a major cause of them. I know of no way in which an area of land could suddenly sink by tens of feet, except by the release of large amounts of gases whose pressure had previously held open a large total volume of pore spaces in the underlying rocks.

Earthquakes in Italy and in Greece are fairly common, and therefore, there are many references to them in the classical literature of Greece and Rome. Volcanoes and earthquakes were the only sources of information about the deeper ground of the Earth. What was down there was clearly rather terrifying, and for this reason alone, these phenomena attracted a lot of attention.

The eyewitness descriptions from antiquity frequently repeat the same themes. Rumblings and hissing noises during earthquakes, sulfurous fumes, changes in ground water levels, hot gases, and flames recur in many of the descriptions. Isaac Newton also subscribed to the view that earthquakes were connected with gases. He wrote that "sulfurous streams abound in the bowels of the Earth and ferment with minerals, and sometimes take fire with a sudden coruscation and explosion, and if pent up in subterraneous caverns, burst the caverns with a great shaking of the Earth, as in springing of a mine."

Tsunamis, Earthquake Spots, and Mud Volcanoes

The same consideration applies to several other phenomena related to earthquakes. The large and sometimes devastating ocean waves called tsunamis are one of those. A rapid and very large change in some volume is necessary to set up these waves, and that volumetric change has to be of a magnitude similar to the volume of ocean water that has been displaced to make either the negative or the positive phase of the great wave. Again sinking of an area of ocean floor due to the sudden escape of gases would be a possibility as would the rapid expansion of gases that make their way from the ocean floor to the surface. There are many reports of violent bubbling of areas of the ocean, and of flames emerging out of the water.

Another feature of earthquakes that seems incompatible with the theory of sheer strain is when the rocks reach breaking point. Earthquakes are known at depths down to 700 km, and the pressure there is so great that sudden fracture cannot occur. The friction between two masses that slide against each other would be so great that this would far exceed any mechanical breaking strength of any rock. Any movement at such depths would occur only as a gradual adjustment proceeding in step with the driving force that causes the movement. This implies that another process must be going on down there and finding the answer to that may also then explain the features of shallower earthquakes that have so far remained unexplained, but that appear in seismic investigation quite similar to the deep ones.

On June 8, 1994, a very large earthquake registering 8.2 emanated from 600 km below Bolivia. There was also a powerful deep earthquake in Colombia in 1970.

While strain measurements in rocks have been carried out in the United States with instruments of the highest accuracy, with the hope that a predictive feature will be found, such measurements have not shown what had been expected. Strain measurements have shown no predictive value, and yet they have remained the mainstay of US earthquake research, almost to the exclusion of any other approach. In California, the major earthquake line is the San Andreas Fault, and it is clear that a shift of one side against the other has occurred repeatedly. This seems to be the reason for the concentration on strain measurements. If, however, one were to start in some other area, one would develop a quite different outlook.

There are other features of earthquakes that also have to be considered. There are places that are distinctive "earthquake spots." There is a spot in northern Norway where for a long time one could almost be guaranteed to feel an earthquake in any 24-h period. These were weak earthquakes, not much above the level at which one could feel them, but there was no fault line that was slipping, no accumulation of any deformation of the surface, it just kept shaking in an area that was about 12 km across. A very similar story comes from two places in the United States, one is on the western tip of Flathead Lake in Montana, and the other is in Arkansas, near the small town of Enola. Both of these have been active in recent times, and the one in Arkansas is known to have been active already some 80 years ago.

Another earthquake spot is on the north shore of the St. Lawrence River, most interestingly just in a large meteorite impact structure or astrobleme called

Charlevoix. The large meteorite struck there some 350 million years ago, and detailed evidence of this impact has been obtained. Despite the length of time that has elapsed since then, it seems that even now the area has not settled down and some activity is still clearly centered there. Some earthquakes that can be felt occur there every few years, and microquakes are registered extremely frequently. In this case, the proximity to the major fault line of the St. Lawrence River complicates the discussion somewhat, but nevertheless, the concentration of the seismic activity to the 30-mile-diameter impact area is quite evident.

Such spots clearly need a different explanation from that of plates shearing against each other. Possibly, the explanation has to do with gases forcing their way up and causing fractures in the rock to open and shut repeatedly.

Soter and I investigated in some detail the Arkansas and the Charlevoix spots, and in the course of this discovered that they both contain a most intriguing feature which has shed further light on this type of occurrence. This is the presence of clusters of earth mounds that stand abruptly out of the alluvial plane. From a few feet to 40 ft in height and up to 200 ft or so in the horizontal dimensions, they are composed internally just of the clay and sand of the local alluvium, and no good reason has been offered to account for their origin. In the Charlevoix region, there are three clusters of such mounds, all in the alluvial fill in the ringed depression that outlines the impact feature. In each case, they lie close to or even directly over known. There are no signs that any rock has been greatly heated. The association of these strange mounds with locally concentrated seismic activity cannot reasonably be ascribed to chance. While such mounds do occur elsewhere, dense clusters of them are extremely rare, and an explanation for them is required. One cannot argue that the shaking of the ground of the earthquakes would itself cause what appears to be a substantial extrusion from below.

A class of a much larger type of feature is known and referred to as "mud volcanoes" which are strongly related to earthquake activity. Mud volcanoes are mountains that are in the general shape of a volcano, sometimes, but not always, with an open hole on top and with steep sides sloping down to the plane below. The sides are made of rock debris, which presumably was ejected at the top as a mixture of such debris with water. Huge fields of mud volcanoes exist in several areas of the globe. The best known ones and the largest are in Azerbaijan on the north slopes of the Caucasus. Large eruptions of individual mud volcanoes are common there, and the gases that propel the eruption are usually flammable and become ignited at the time, presumably by electrostatic sparks resulting from the friction of fast-moving rock grains. Flames to a height of 2 km have been photographed from one mud volcano whose orifice measures 120 m across.

The gases coming out of mud volcanoes have often quite unusual composition and contain elements that are known to be at a high concentration in the mantle of the Earth and at a much lower concentration in the sediments and in the outer crust. They clearly represent a very different chemical environment from that of the sedimentary cover.

The mounds on the earthquake spots in Enola, Arkansas, and Charlevoix on the St. Lawrence River can be attributed to the same class of phenomenon as mud volcanoes, only on a much smaller scale. Some have clearly visible holes on top and some even show that the ground has deformed in recent times so that the trees that are now there all lean outward from the axis. The mounds at Charlevoix must be younger than 10,000 years, since they could not have survived the last ice age with a large glacier in that position. The spot in Enola similarly has a dense field of these features, also overlying the spot on which the earthquakes are centered.

My opinion is that high-pressure gases erupting from deep levels seem to be responsible for these features and presumably therefore also for the small earthquakes. Expulsions of high-pressure gas from deep levels will throw up a mixture of gas and mud, and the mud will flow down the mountainside and dry up, thus creating the hill of the general shape of a volcano.

Further Evidence for Massive Gas Emissions

Massive emissions of natural gas from the ground have been observed on many occasions and in many locations. One set of observations is that referred to as "mystery clouds." This is a cloud phenomenon which takes the form of a funnel-shaped cloud emanating from a small spot on the surface—land or water—and then rising and spreading out with increasing height, as well as showing increasing deflection from the vertical. Mystery clouds have reached heights of 30,000–40,000 ft, the upper limits of commercial aircraft heights. Two hundred sightings of these clouds have been recorded, mostly by satellite photography, but some also from high-flying aircraft. Satellite measurements of temperatures within the funnels found temperatures to be some 30°C lower than the surrounding air at the same height. The only possibility for creating a rising column that is much colder than the air around it is to make it from a gas or a gas mix that is lighter than air (mean molecular weight of air $\approx$30, the molecular weight of methane is 16). A gas lighter than air will rapidly move upward, and its expansion will make it cool very rapidly. These observations prove that the gas involved is indeed much lighter than air and the only prominent gases that are in the rocks that could be held responsible are hydrogen and hydrocarbon gases; both of these gases are combustible.

It is my view that these mystery clouds represent sudden violent emissions from the rocks both on land and from the sea floor. As such, a gas mass emerges into the air, the rapid cooling will generally cause condensation along the contact with the outside air, and it is this which makes it visible as a cloud. Many satellite pictures show these funnel clouds very clearly, either seen directly or their shadows are seen very prominently in arctic latitudes on white snow or ice fields. A methane–air mix in a combustible mixing ratio with 20% air would still be 35°C colder but could still reach up to 40,000 ft.

The magnitudes of these gas emissions are very large, as judged not only by the size of the clouds but also by the amounts of seawater they sometimes contaminate. Hydrocarbon gases in the water will make it anoxic, as microorganisms will quickly use up the dissolved oxygen to derive their energy supply from the oxidation of the hydrocarbons. This alone will be damaging to fishes. But more damaging still will be the hydrogen sulfide produced as the microbial activity obtains its oxygen from sulfates in the seawater. Hydrogen sulfide makes the water milky and is extremely toxic to most sea fauna. Fish kills are reported from many parts of the world, and often from areas that are known to be hydrocarbon rich. In Japan, such fish kills are often reported both before and after earthquakes. The fishermen frequently report their surprise at seeing many bottom dwelling dead fishes floating on the top. On the coast of Namibia, along the west coast of South Africa, such events were quite frequent, with the milky hydrogen sulfide water, but these reports were thought to represent minor local events. But then on March 18, 2001, a satellite image captured the discoloration of the sea, stretching over a distance of more than 200 km. A second emission took place 16 days later, and the discolored water drifted and covered an area of 20,000 km^2. Whether a massive mystery cloud was observed at the time is not known. But the atmospheric effects are much shorter lived than the oceanic.

In addition to the mystery clouds, we have information of the changes that massive amounts of gas have caused on the mud of the ocean floors. There are large areas where circular markings between 1 m and 100 m in diameter are seen by sonar or visually, and they are usually associated with an excess of methane in the water even long after the violent eruption has ceased. These "pockmarks" were first discovered by M. Hovland in the North Sea, overlying the large gas fields that are known to be there. They have since been found also in many areas around the globe, most frequently near the margins of the continents and above or near gas-producing areas.

How do we know that these pockmarks are related to earthquakes? Sonar can detect features buried in uniform ocean mud. In some instances, pockmarks have been seen densely spaced on the surface of the mud, suggesting that they were caused by the same event, but then in the same region, another deeper horizon was found with another field of pockmarks again all at a common level. From estimates of the rate of deposition of the mud in the area, it was judged that this event preceded the other by about 1,000 years.

With so much information about the eruptions of hydrocarbon gases, we may wonder whether the supply of carbon to the atmosphere is due in major part to these eruptions and not due to the emission of CO_2 from volcanoes. We know that the deposition of carbonate rocks would diminish the atmospheric oceanic CO_2 to the extent that plant life would be diminished or disappear in a time of the order of one million years, and we must therefore look for continuous supplies of carbon from the interior. It would not matter much whether the carbon-bearing fluids were hydrocarbons or CO_2 since in our oxidizing atmosphere, hydrocarbon gases would be quickly turned into CO_2.

It has sometimes been argued that one can make that distinction from observations of the carbon isotope ratio which seems to have remained almost constant over most of geologic time. That is seen in the carbonate records that were mostly derived from atmospheric CO_2 which at any time would be well mixed globally; whether before mixing in the atmosphere the carbon isotope ratio of each source of supply would have been constant, we do not know. It is only the average value that we can observe in the carbonate record. If hydrocarbon gases had been responsible for that carbon supply, it is only the average value over long times that we can identify, but individual outbursts could have varied greatly from the mean. We can in fact see great variations in the hydrocarbon gases that come up from the ground in different locations and at different times. The fact that random drilling and all deep drilling show hydrocarbon gases and only rarely CO_2 would then make it more probable that the main carbon supply into the atmosphere has come from hydrocarbons.

Are Earthquakes Caused by Gas Eruptions?

How would we explain earthquakes in terms of gas eruptions? How would all the known effects fit together? Those are the issues I address in this section.

For the earthquake spots, repeated gas eruptions would seem to be the only possible explanation. They are simply locations where a puff of gas comes through the ground frequently, and the seismic disturbance is simply due to pore spaces being opened and closed each time. The fact that they are frequently associated with mounds in the surface soil suggests that the gas events have occurred sufficiently frequently to have expelled the amounts of alluvial materials to create the mounds.

Large vertical displacements of areas of land can be understood if a mass of gas had previously held open pore spaces in the rocks below, and thereby raised the ground, and if these pore spaces had suddenly made connections to the surface and rapidly exhausted the gas. Such volumetric changes occurring in a matter of seconds can then account for the large tsunamis and for the flames often seen in earthquakes. As methane appears to be the most common gas in the rocks, it would seem reasonable to expect that methane would be the principal gas responsible, just as it is known in the case of mud volcanoes and pockmarks. The mud volcanoes show locations in which earthquakes are particularly frequent and locations in which large amounts of underground mud have been generated by the frequent agitation of groundwater in fine-grained alluvial sediments.

With the strain measurements having no predictive value, would it not perhaps be better to concentrate on measurements of gas pressure in the pore spaces in the rocks? I asked one of the leading seismologists in California whether he thought this was a good plan, since, after all, the only successful prediction, avoiding a large loss of life, at Haicheng, was based entirely on gas effects. The reply I got was that there was not enough money in the budget to pursue any other avenue of research. I said I could not believe that, but if he would like, I would myself pay for one valuable line of research. He looked perplexed. Was I perhaps independently very wealthy, and did I offer to pay for a research program on the scale to which he was accustomed? Like building tunnels through mountains to lay vacuum pipes with laser beams in them to find displacements of one millimeter. He could not readily turn down such an opportunity, so he replied: "Of course, what do you propose?" My reply was that I would buy 20 or 30 bottles that have a good airtight closure top, fill them with carefully distilled water, and distribute them among seismic experts in California. I would ask each one to have one bottle at home, one in his car, and one in his lab. When an earthquake occurs of sufficient strength so that one can feel it, he should then merely open the bottle, pour the water out, and close it firmly. A careful analysis of the gas at a later time would give information about the types of gas and approximate information about the quantities involved that had contaminated the air at the time.

The reply I received was, "And what would that tell us?" I replied that I thought it would be better to make this simple observation first, and then decide on an explanation. What if it contained a high helium concentration? Or some other gases that were rare in the atmosphere?

In fact, he turned down my offer, and when the earthquake struck at Loma Prieta on October 17, 1989, shortly after this conversation, no gas samples had been taken for analysis. But in fact, there were two observations before the earthquake at Loma Prieta on October 17, 1989 that seemed to be gas-related and are clearly just prior to the earthquake. These observations were made for different purposes, unrelated to earthquake research, and yet they constitute the best earthquake-predictive observations that seem to have been made at the time. One was the observation of the amount of helium in a shallow well, which showed a sharp increase before the quake. The observation was made a day before the quake, but that day was chosen just on the regular weekly schedule for visiting the site. The helium increase may well have commenced earlier, it was certainly a very large increase.

An increase in the concentration of helium does not just indicate that other gases have entered the region, but it specifically indicates that the gases have originated at lower levels. As I have stressed before, there is no other way in which a concentration of helium could arise since helium cannot be concentrated by any chemical action, and it will certainly not migrate to one location horizontally or even downward. And no other explanation for a sharp helium increase at one level is known other than the forcing up of diffusely distributed helium by another upwelling fluid. The fluids responsible for the earthquake would then be supposed also to have come from below.

The other observation of Fraser-Smith was that of a low radio frequency noise that is not normally present. I attribute this to the interruption and reconnection of earth currents flowing in the groundwater, as these current paths are interrupted and reconnected by the bubbles of insulating gases that stream through the pores of the rock. Would these and other gas-related precursory effects not form the best line of earthquake investigation, to devise the most important of all, a predictive capability?

Can Earthquakes Be Predicted?

We see that the eyewitness descriptions make major earthquakes look much like violent eruptions, quite similar to gas eruptions from volcanoes or mud volcanoes. The airborne noises, the flames, the air pollution are all similar, and while most of the intense effects take place at the time of the quake, some of the effects occur as precursors and cannot therefore be ascribed to secondary effects of the mechanical deformation of the ground. It seems very strange to me that in all the attempts to predict earthquakes, in the U SA, no gas observations are included. Highly accurate measurements of the distortion of the ground represent the main effort, since the current theory has earthquakes resulting only from a gradually augmenting stress in the rocks until they reach the breaking strain, which then represents the earthquake. It is therefore supposed that one can measure the building up of the stress by the slight deformation prior to a quake. However, as a means to predicting earthquakes, this method has been entirely unsuccessful. The ground does distort on occasions, but not by any unusual amount before an earthquake.

Gases can indeed have a lot to do with earthquakes. A large volume of gas entering the crust of the Earth from deeper levels and at a high pressure will greatly change the mechanical properties of the rock. Pore spaces will be inflated, and the overburden weight of the rock will be effectively relieved by the pressure of the gas. The great weight of the overburden would normally have resulted in high internal friction, opposing any slippage at all but the shallowest levels. But with gas effectively bearing the overburden, slippage can occur much more easily. Much smaller values of stress in the rock will then be sufficient to cause a quake.

The sudden change of the mechanical properties has often been observed, but has been ascribed to the shaking of the ground, leading to a condition that has been named "fluidization." There is no demonstration of such a process, except in earthquakes. The word is invented because it happens, and once a process is given a name, it is often assumed that it is independently known and demonstrated. Of course, the flow of high-pressure gas through alluvial soil will separate the particles and give a mass of them a fluid-like behavior.

The absence of high stresses along the St. Andreas fault was indeed a surprise to the investigators, when they had a chance to make such measurements in the deep well drilled at Cajon Pass in Southern California. They failed to find there the extra heat that the known past slippage should have left behind, had it taken place without gas levitation, as well as the removal of extra heat by the streaming gas.

When gas has invaded an area of the crust, it generally shows some emission at the surface that can be observed, and that results in the various effects mentioned. Of course, the gases that were in the pore spaces, to start with, are pushed up first, before the "new" gas has got to the surface. This brings up smells which cause surprise or consternation among many animals; it brings up more carbon dioxide and less oxygen than air has normally, and this drives animals out of burrows; it brings up humidity and temperature of the subsurface and thus frequently makes a fog. This contains more of the heavy CO_2 molecule than the average air and can therefore make a warmer cloud that stays on the ground instead of rising rapidly. Radioactive gases that are normally generated in the ground make a prominent appearance as they are flushed from the ground.

These signs should be taken to mean that the rock underneath has now suddenly lost much of its strength, and even small stresses will allow it to break. There was no particular buildup of stress prior to the quake, and measurements of this are therefore useless as predictors. The sudden event was the gas invasion that weakened the rock, and it is on this that a prediction method has to be based. During earthquakes and after, a lot more gas escape can usually be observed, and by then, the deep source gas may have made its way to the surface. This is often combustible, probably mainly methane as this is in most common gas in deep rocks, and it often catches fire.

In China, in Japan, and in the Soviet Union, much more attention was paid to gas phenomena. Japan even has a "Laboratory of Earthquake Chemistry." The USA is far behind in this field, not because it does not have the technology, but just because it took a wrong choice some time ago and now does not wish to change course. But the citizens of earthquake-prone regions will be more concerned with obtaining a warning than to be party to a scientific controversy. Subsurface gas observations are simple and comparatively inexpensive, such as changes in groundwater levels in water wells or changes in gas pressure above a water table. It is high time that California and the central Mississippi region obtained the knowledge and experience in this field that will be necessary to establish a meaningful prediction service. Instrumentation operated by scientists is one aspect of this; public earthquake education and a reporting network are another, to ensure the widest possible coverage for the observation of the many phenomena that may be relevant for predictions. One wonders how many such observations go unreported because their relation to earthquakes is not generally known.

How massive and extensive such gas emissions can be was seen by satellite observations of the Namibian coast of Southwest Africa. It is a stretch of coastline known for its commercial production of natural gas. People living along the coast had observed massive fish kill events occurring several times a year, but had thought of the events as being localized and not of the magnitude displayed on the satellite photography. The gas entering the water contains hydrogen sulfide, a substance highly poisonous to fish and with the great advantage for observations of discoloring the water close to the surface. It was satellite color photography on March 18, 2001 that first revealed the enormous extent of the phenomenon.

The same event had discolored the water for a distance along the coast of 200 km. The concentration of oxygen from the water over this stretch was much below normal, as is commonly the case in methane-emitting regions. While most of the emissions that lead to the funnel-shaped mystery clouds come from a concentrated small spot, this event originating over such a long line must also have had a deep origin since one cannot explain the simultaneity of the emission over such a large distance if it had derived from shallow sources. The discolored water was still seen more than 2 weeks later, having drifted with the prevailing water drift and having by then covered a surface area of 20,000 km^2. Several more such events have been observed by satellite photography since then.

It is very probable that methane is the principal component of the gases. Firstly because the anoxia of the water requires a much more abundant oxygen sink than hydrogen sulfide could possibly provide and secondly because we know of the whole of this coastline having a subsurface content composed mainly of methane.

Do the Large Emissions of a Light and Flammable Gas Pose a Hazard for Aircraft?

One would certainly think so, if one recognized the magnitudes of the emissions that are on record, both for the mystery clouds, the mud volcanoes, and the pockmarks and other gas-related events. Flashes of flames have been seen going up from the ground to high aircraft altitudes, and if an aircraft had been present there, it would certainly have been troubled by such an event. I have close acquaintance with one gas-related case in 1982 where a Boeing 747-200 came very close to disaster, but was able to save itself. British Airways and Rolls Royce commissioned me to make a study of the event, which included detailed interrogations of the captain and crew, as well as access to the physical data that had been obtained after the successful landing. Two remarkably similar events with the same type of aircraft occurred later, so that about 750 persons narrowly escaped disaster.

The flight I investigated was at night, at a height of 37,000 ft, over the island of Java, in an area in which some volcanic activity had been observed. Suddenly, bright specs appeared on the windshield and soon changed into bright lines running up its whole length. Then, within a minute or two, all four engines stopped. Captain Moody, the flight captain, had been in the passenger cabin and returned quickly to the flight deck. He put the aircraft into its best gliding attitude and gave instructions to the flight crew to attempt a restart of the engines. Several attempts failed during the glide down to 15,000 ft. By then, he was close to mountain altitudes, and he steered the aircraft out toward the sea. He gave emergency instructions to the passengers to brace for ditching at sea. Meanwhile, the bright lines continued and had spread to the front surfaces of the wings and especially to the engine air intake housings, from where a bright beam of light now emerged, illuminating what appeared to be a tenuous cloud ahead (Figs. 11.1 and 11.2).

What was all that? What could cause these strongly luminous particles to appear? How were they related to the stopping of all engines? It was not a phenomenon the captain and crew had experienced before or had ever heard about. It was a terrifying situation, since the aircraft could not support itself for more than 25 min without engine power, and that could not bring it to an airport.

Then the bright specs suddenly disappeared. On the next restart attempt, all four engines started by turns, and Captain Moody guided the 747 back to Jakarta, the nearest airport, where it landed successfully. It was clear that the airframe had suffered some damage. The windshield was so heavily abraded that the landing was performed without the flight deck crew seeing more of the airport than the glare of the lights lining the runway. Also one engine had run unevenly and had been shut off.

So what had caused all that? Volcanic ash presumably was the source of the small particles. But how did they get lifted up to 37,000 ft? How did they become so luminous? Did they stop the engines? How?

Firstly, it is quite clear that small particles could not have reached the altitude, except by being lifted up together with a gas. It is also clear that the particles had to

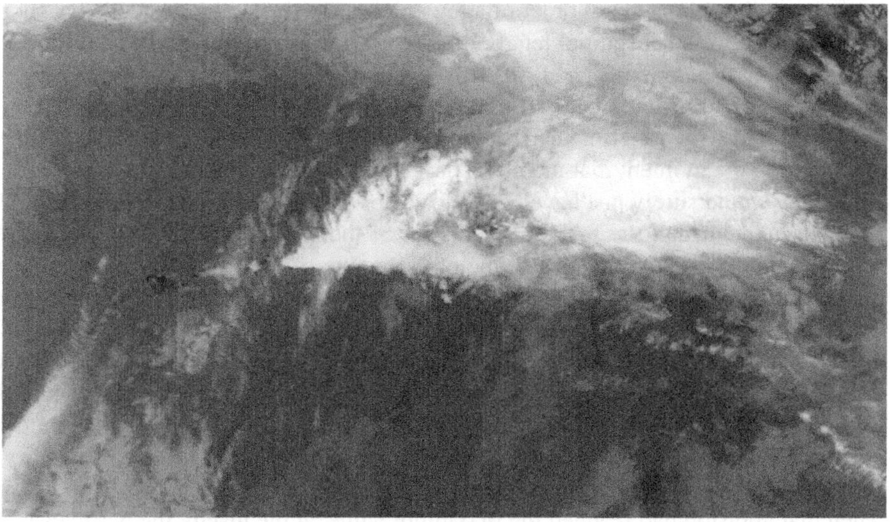

Fig. 11.1 Mystery cloud at night illuminated by low sunlight

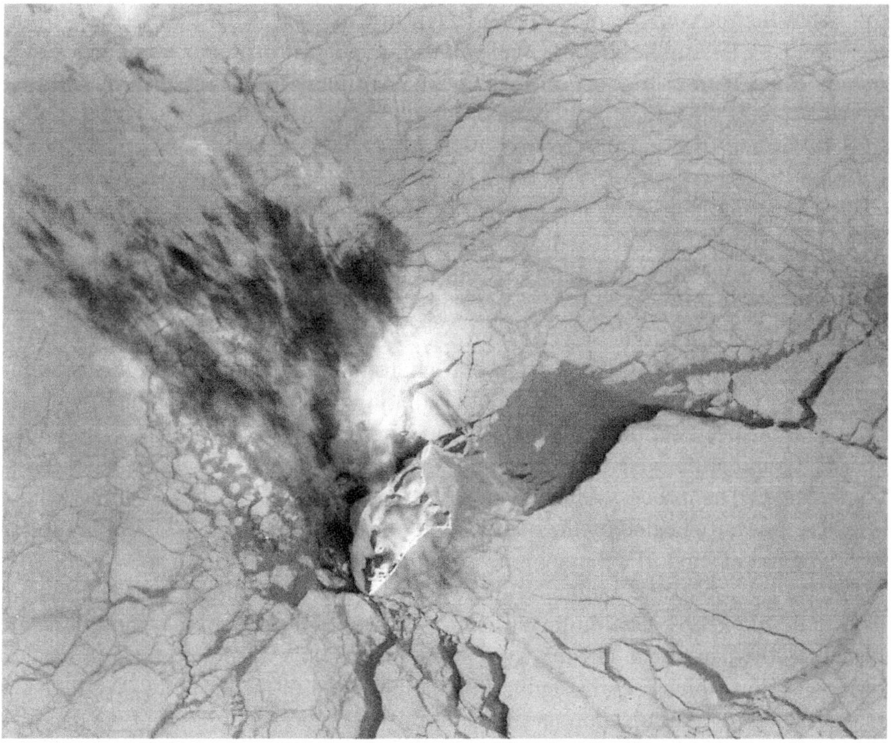

Fig. 11.2 Mystery cloud seen in daylight as a shadow projected on an initially smooth ice field

be very small, of the general order of a few microns in size, to be suspended at such high altitudes for a substantial length of time. What did small particles have to do with stopping the engines? The Rolls Royce engineers who investigated the engines said that they had suffered badly from abrasion, and they were not surprised that the engines failed to work. But that leaves two questions unanswered: Why did the engines stop the moment that the plane flew into the cloud containing the particles, when they would surely just have begun to suffer abrasion, and yet why did they all start the moment that the plane traveled out of the cloud? At that time, they had suffered the maximum abrasion. Some other feature of the cloud has to have been responsible for the engine stoppage. A feature that disappeared as the aircraft left the cloud.

The passengers in the aft section of the cabin all stated that they saw several times long flames coming from the engine exhausts. Presumably, this occurred during the restart attempts that failed. For both these reasons, we have to look to a chemical explanation. Heavy abrasion by micron-sized particles is not a possibility, for if it had caused engine stoppage, it would not allow all engines to start again when that abrasion had reached the maximum value, at the instant the particles had disappeared. The only explanation that I could see was that the gas of which the cloud was composed was a mixture of air with a light hydrocarbon gas, most likely methane. The engine stoppage would then be caused by the engines receiving much too rich a fuel mixture, namely, the injected fuel as well as the fuel resident in the gas cloud. A jet engine given a fuel mixture too rich for proper combustion will indeed cause flames to come out of the exhaust in any restart attempt. It will also mean that once the aircraft is outside the cloud, it will again have the correct mixing ratio of air and fuel and the engines evidently started despite the erosion they had suffered.

But how did the particles cause the bright light and high values of erosion? The chemical energy content of the small particles can be calculated for the most favorable composition and that turns out to be much too small to cause streaks of light going over the entire windshield. They would have to be glowing brightly for a time of at least 1 ms to achieve this effect.

In combustion engineering, an effect where the circumstances are not suitable for a flame propagation but where a hot radiating particle can accelerate the flame propagation speed and thus maintain a flame around the particle even if streaming speeds would otherwise have swept the flame away is known as smoldering combustion. The process is well known from the example of a candle whose wick is constantly heated by the combustion of the vapor arising in a pool of molten wax which will support a flame in the presence of high evaporation speeds only if it is held to the top of the candle by the wick which keeps being heated, and whose heat radiation keeps igniting the streaming vapor. In a similar way, the grains are heated by the flame that surrounds them, the flame which is due to the combustion of the gas, but which remains ignited by the radiation of the particles. We then see that the glowing particles are an indication of a combustible gas, and we can also understand that at the white heat to which the particles are raised would cause heavy erosion despite being so small.

The examination after the incident, of the material accumulated in various air intake passages, showed the presence of a few percent of unoxidized carbon particles, which implies that subsequent to leaving the volcanic temperatures, they were cooled by the expansion of the gas with which they came, before they could be oxidized by mixing with air. The interrogation of the flight crew yielded the answers given independently by all members that a strange smell was perceived at the time when the luminous flakes were seen, a smell they all described as the smell of burnt oil or the smell of the London Underground, which is indeed due to burnt oil that drips on the conductor rails from the lubrication of the machinery.

The sequence of the events is then accounted for. It must have been a likely sequence since exactly the same phenomena were encountered over Java a few days later by a Singapore Air 747 and again in another incident many years later, by a KLM 747 over the Aleutian Islands.

Since all these incidents occurred at night, we have no knowledge whether mystery clouds existed at the time. But at least we have three detailed accounts of very serious incidents whose daytime equivalents we can only guess. Whether the relation to volcanic areas is significant is not known, since such gas emissions take place over many nonvolcanic areas also, and the force of a volcanic eruption has little effect for determining the height to which a cloud would rise. The presence of volcanic small grains may be significant for generating the bright specks, but not for stopping the engines, which was the main hazard. These incidents should be considered closely in the investigation of any aircraft disasters for which no clear and unique explanation has been found.

What are the effects that a cloud lighter than air and possibly a flammable mix would have on an aircraft? Firstly, a lot of instruments will give erratic and false readings. The air speed indicator most likely will give too low a reading. The barometric height indicator will give too high a reading as there is less weight of gas above it. The air flow around the wings may show them to be near stalling speed since the aircraft was not designed to be supported by a lower density gas. All these effects will be confusing to the pilots, especially if they have never contemplated that such effects could occur. They will be equally confusing to the autopilot that was also designed to operate in plain air. The confusion caused by these false readings may itself become the cause of an accident.

The chemical changes probably represent the most serious threats as they did in the three "near accidents" we have discussed. The engines may stop operating because they have a wrong fuel–air mixture. Flames may be ignited especially in confined spaces such as cable ducts that carry control cables, since flames can be maintained much more readily where they are shielded from the fast air flow around the aircraft. The engines may ignite a large external mass of flammable gas initially close to the plane, but eventually, a large cloud that is mainly above the plane as the gas is rising at a high rate. Large external explosions of methane may go unidentified in daytime as a methane–air mix is rather transparent and its flame does not make much light. For the investigators of an accident, the absence of shrapnel holes in the skin of the plane will seem to deny that a massive explosion was the cause of the disaster. The slow flame propagation speed in a methane–air mix will mean that very little noise is radiated from a small cloud that has been ignited, and only when a cloud is large enough so that the sum of the expansion speed of the burning gas and the flame propagation speed reaches the sonic velocity a loud booming sound will result.

Another feature of accidents due to rising gas clouds is that they have a strong geographical link. Regions that are known to be commercial gas producers will send up clouds more often than average regions. The edges of continental shelves are prone to be gas rich as we have seen in many cases where commercial extraction of gas takes place but also from the presence of pockmark fields on the margins of continents and from the presence of the water–methane ices called methane hydrates that are found lining many continental shelves. The continental shelf edges northward of New York City represent an area in which many more accidents have happened than would be statistically expected. My proposal to the National Transportation Safety Board (US) to survey the region for pockmarks, a very simple and inexpensive procedure, was declined. However, much evidence is acquired about massive gas eruptions having occurred, they are regarded as impossible.

I tried to alert the International Airline Pilots Association and asked them to enforce the following safety rule: to prevent a recurrence of the types of events that had occurred in the volcanic regions. My proposed rule was that if you see the bright specks or streaks you must make a 180° turn since that would be the only direction of which you knew that it would quickly take it out of the cloud. But the problem was misunderstood not only by that Association, but apparently by most of the scientists with whom they consulted on this subject. It seems that they all thought that it was specifically a volcanic event, and they did not understand that it must have been a massive gas emission event that may well have come from volcanoes. The main threat to the aircraft that had arisen, and had stopped the engines in each of those cases, was not due to the presence of volcanic ash, but rather that its presence had made the gas event visible to the flight crew. It seems that the general thinking was that the particles could only be propelled to this altitude by the force of a volcanic explosion. It was not recognized that a gas lighter than air was the only way in which small particles could be lifted to 37,000 ft. Not even a heavy machine gun bullet can reach such altitudes and the force of a volcanic explosion could certainly not have done this. Small grains cannot be made to shoot through that distance in air by any explosion. The association therefore adopted only the rule that pilots should be informed of volcanic eruptions in the region over which they were flying.

At the time of writing, this is still the only rule adopted, despite the hundreds of mystery clouds that have been seen in nonvolcanic areas, but have similarly reached such high altitudes. All accident investigations have consistently refused to consider that gas eruptions in nonvolcanic areas could lead to disasters and have therefore attributed many disasters to arbitrarily selected causes. They have thereby dismissed the valuable evidence that could be obtained only from flight crews that had narrowly escaped a gas emission disaster.

Bibliography

Agnew, D.C.: The 1952 Fort Yuma earthquake—two additional accounts. Seismol. Soc. Am. Bull. **68**, 1761–1762 (1978)

Aguilera, J.G.: The Sonora Earthquake of 1887. Seismol. Soc. Am. Bull. **10**, 31–44 (1920)

Alippi, T.: Nuovo contributo all'inchiesta sui "brontidi". Boll. Soc. Sismol. Ital. **15**, 65–77 (1911)

Ali-Zade, A.A., Shnyukov, E.F., Grigoryants, B.V., Aliyev, A.A., Rakhmanov, R.R.: Geotectonic conditions of mud volcano manifestation in the world and their role in prediction of gas and oil content in the Earth's interior. Proc. 27th Int. Geol. Congr. **13**, 377–393 (1984)

Bagnold, T.: Extraordinary effect of an Earthquake at Lima, 1828. Quart. J. Sci. Lit. Art **27**, 429–430 (1829)

Bischof, G.: On the natural history of volcanoes and earthquakes. Am. J. Sci. **37**, 41–77 (1839)

Buskirk, R.E., Frohlich, C., Latham, G.V.: Unusual animal behavior before earthquakes: a review of possible sensory mechanisms. Rev. Geophys. Space Phys. **19**, 247–270 (1981)

Degens, E.T., von Herzan, R.P., Wong, H.-K., Deuser, W.G., Jannasch, H.W.: Lake Kivu: structure, chemistry and biology of an East African Rift lake. Geol. Rundschau **62**, 245–277 (1973)

Demetrescu, G., Petrescu, G.: Sur les phenomenes lumineux qui ont accompagne le tremblement de terre de Roumanie de 10 Novembre 1940. Acad. Roumaine Bull. Sec. Sci. **23**, 292–296 (1941)

Deng, Q., Jiang, P., Jones, L.M., Molnar, P.: A preliminary analysis of reported changes in ground water and anomalous animal behavior before the 4 February 1975 Haicheng earthquake. In: Simpson, D.W., Richards, P.G. (eds.) Earthquake Prediction: An International Review, pp. 543–565. American Geophysical Union, Washington, D.C (1982)

Donovan, T.J., Friedman, I., Gleason, J.D.: Recognition of petroleum bearing traps by unusual isotopic compositions of carbonate-cemented surface rocks. Geology **2**, 351–354 (1974)

Fitz-Roy, R.: Sketch of the surveying voyages of his majesty's ships Adventure and Beagle, 1825-1836. J. Roy. Geogr. Soc. **6**, 311–343 (1836)

Fuller, M.L.: The New Madrid Earthquake, p. 494. U.S. Geological Survey Bulletin, Washington DC (1912)

Galli, I.: Raccolta e classificazione di fenomeni luminosi osservati nei terremoti. Bol. Soc. Sismol. Ital. **14**, 221–447 (1911)

Gold, T.: Terrestrial sources of carbon and earthquake outgassing. J. Petrol. Geol. **I**(3), 3–19 (1979)

Gold, T., Soter, S.: Brontides: natural explosive noises. Science **204**, 371–375 (1979)

Gold, T., Soter, S.: Natural explosive noises. Science **212**, 1297–1298 (1981)

Gold, T., Soter, S.: Methane and seismicity: a reply. EOS **64**, 522 (1983)

Gold, T., Soter, S.: Fluid ascent through the solid lithosphere and its relation to earthquakes. Pure Appl. Geophys. **122**, 492–530 (1985)

Goodfellow, G.E.: The Sonora Earthquake. Science **II**, 162–166 (1888)

Hauksson, E.: Radon content of groundwater as an earthquake precursor: evaluation of worldwide data and physical basis. J. Geophys. Res. **86**, 9397–9410 (1981)

Haywood, J.: The Natural and Aboriginal History of Tennessee. Nashville, Tenn (1823)

von Hentig, H.: Animal and earthquake. J. Comp. Psychol. **3**, 61–71 (1923)

Higgins, G.E., Saunders, J.B.: Mud volcanoes—their nature and origin. Verhandl. Naturf. Ges. Basel **84**, 101–152 (1974)

Hovland, M., Judd, A.G., King, L.H.: Characteristic features of pockmarks on the North Sea floor and Scotian Shelf. Sedimentology **31**, 471–480 (1984)

Humboldt, A., von: Personal Narrative of Travels to the Equinoctial Regions of the New Continent During the Years 1799-1780, vol. 2, 3rd edn., pp. 211–212. G. Bell, London (1822)

Ippolito, F.: An account of the earthquake which happened in Calabria, March 28, 1783. Philos. Trans. Roy. Soc. **73,** Appendix i–vii (1793)

Kanamori, H., Ekstrom, G., Dziewonski, A., Barker, J.S.: An anomalous seismic event near Tori Shima, Japan: a possible magma injection event. EOS **67**, 1117 (1986)
Lacroix, A.: La Montagne Pelee et ses Eruptions 194–195 (1904)
Lalemant, H.: Relation of what occurred most remarkable in the Missions of the Fathers of the Society of Jesus in New France in the years 1662 and 1663. In: Thwaites, R.G. (ed.) The Jesuit Relations and the Allied Documents, vol. 48, pp. 17–179. Jesuit Missions, New France (1663)
Larkin, E.L.: The great San Francisco earthquake. Open Court **20**, 393–406 (1906)
Lawson, A.C., et al.: The California earthquake of April 18, 1906. Carnegie Institution, Washington, D.C (1908)
Lee, W.H.K., Ando, M., Kautz, W.H.: A summary of the literature on unusual animal behavior before earthquakes. U.S. Geol. Survey Open-File Report, 76–826 (1976)
Le Guern, F., Tazieff, H., Fairre Pierret, R.: An example of health hazard: people killed by gas during a phreatic eruption: Dieng Plateau (Java, Indonesia), February 20th, 1979. Bull. Volcanol. **45**(2), 153–156 (1982)
Li, J.: Earthquake – a harvest of agony. Los Angeles Times, Oct., 1980
Liao-ling Province Meteorological Station: The extraordinary phenomena in weather observed before the February 1975 Hai-cheng earthquake. Acta Geophys. Sin. **20**, 270–275 (1977)
Michell, J.: Conjectures concerning the cause, and observations upon the Phenomena of Earthquakes. Philos. Trans. Roy. Soc **51**, 566–634 (1761)
Ninkovich, D., Sparks, R.S.J., Ledbetter, M.T.: The exceptional magnitude and intensity of the Toba eruption, Sumatra: an example of the use of deep-sea tephra layers as a geological tool. Bull. Volcanol. **41**, 286–298 (1978)
Pierce, W.L.: Evening Post, New York, 2 February 1812
Raffles, T.S.: The History of Java, vol. I, p. 28 (1817)
Reitsema, R.H.: Gases of mud volcanoes in the Copper River Basin, Alaska. Geochim. Cosmochim. Acta **43**, 183–187 (1979)
Rethly, A.: A Karpatmedencek Foldrengesei. 445–1918. Academic Publishing House, Budapest (1952)
Schmidt, A., Mack, K.: Das Suddeutesches Erdbeben vom 16 November 1911. *Wurtt, Jahrbucher f. Statist. u. Landeskde.*, Jahrg. 1912, Heft I. 96–139 (1913)
Simon, C.: An account of the earthquake in New France, 1663. In: Thwaites, R. G. (eds.) The Jesuit Relations and Allied Documents, vol. 48, pp. 181–223 (1663)
Sokolov, V.A., Buniat-Zade, Z.A., Goedekian, A.A., Dadashev, F.G.: The origin of gases of mud volcanoes and the regularities of their powerful eruptions. In: Schenck, P.A., Havenaar, I. (eds.) Advances in Organic Geochemistry 1968, pp. 473–484. Pergamon, Oxford (1963)
Stehn, C. E.: The geology and volcanism of the Krakatau group. Proc. Fourth Pacific Science Congress (Batavia), pp. 1–55 (1929) (Reprinted in T. Simkin and R. Fisk (1983), Krakatau 1883, Smithsonian Inst. Press, Washington.)
Stoqueler, Mr: Observations, made at Colares, on the Earthquake at Lisbon, of the 1st of November 1755, by Mr. Stoqueler, Consul of Hamburg. Philos. Trans. Roy. Soc **49**, 413–418 (1756)
Sugisaki, R., Sugiura, T.: Geochemical indicator of tectonic stress resulting in an earthquake in central Japan, 1984. Science **229**, 1261–1262 (1985)
Wallace, R.E., Teng, Ta-Liang: Prediction of the Sungpan-Pingwu Earthquakes, August 1976. Am. Seismol. Soc. Bull. **70**, 1199–1223 (1980)
Williams, S.: Observations and conjectures on the earthquakes of New England. Am. Acad. Arts Sci. Memoirs **I**, 260–311 (1785)
Winslow, C.F.: Earthquake at Lima, March, 1865. Am. J. Sci. **40**, 365 (1865)

Chapter 12
Unfinished Business

It is almost inevitable that an autobiography has to end with a chapter on unfinished business. At no point in one's life can one decide what business can still be finished, and at some stage, the autobiography has to be delivered to the publisher. An autobiography that is mainly concerned with the author's work can never be firmly closed, unless the author has really decided to work no more. I have certainly not done this. So I would at least like to indicate what work I now regard as complete and secure at this stage and what will still suffer debate and possibly modification.

I feel I can safely predict that the theory of hearing, of the function of the inner ear, is here to stay. Of course additions and significant detail will still be added, but the basic notion that the body supplies energy to the hearing apparatus and that this causes the sharp frequency discrimination and the high sensitivity will most likely survive all humanity and perhaps all mammals.

Many details of that theory have by now been worked out, such as the quality of the acoustic match to the incoming sound, the energy transfer in the middle ear, and the velocity of the traveling wave along the membrane on which the detection takes place, and all calculations show these adjustments to be very near to the theoretical optimum. In other words, the design of the ear has evolved to an instrument of great perfection. This in turn certifies that the theory on which the calculations are based was correct.

It was very fortunate that Dr. David Kemp who discovered my 40-year-old paper had also come to the investigation of theories of hearing from a background of radio technology, as I had. He therefore readily understood my reasoning and invented an important medical application based on that theory.

He had demonstrated otoacoustic emission, which is the emission from the ear, of some sounds following an intense stimulation such as a loud click. He could record this emission from a microphone inserted into the ear. I had already mentioned in my papers that we might expect such effects in some cases, but we

---

Editorial note added by Simon Mitton
This chapter titled Unfinished Business was itself an unfinished work in progress when Tommy passed away on June 22, 2004.

Thomas Gold *edited by* Simon Mitton, *Taking the Back off the Watch*,
Astrophysics and Space Science Library 381, DOI 10.1007/978-3-642-27588-3_12,

had not been able to obtain instruments of sufficient sensitivity to observe them. Kemp, at this later time, was indeed able to obtain such instruments. He then discovered an important medical application for such an apparatus. He knew that it was very important to check newborn babies for deafness, as they must then be treated quite differently from babies that can hear. Apparently, the absence of the feedback mechanism is the most common cause of deafness and results, of course, in the absence of this delayed emission. He then produced the appropriate equipment, which is now widely used the world over, just for the purpose of checking the hearing of babies. While no one has yet been able to repair the hearing mechanism if it was deficient in the feedback mechanism, he achieved the next best thing which is to have an objective observation of deafness that would give an early warning of this condition, greatly helping the upbringing of deaf children.

Another area that I have mentioned is the steady-state theory of the universe. I personally still favor it greatly over the big bang theory, which seems to me to contain an arbitrary adjustment to fit each new observational fact, as it is discovered. The steady-state theory is based on a principle that could be shown to be wrong but has so far been seen in agreement with all new observations. The observation of acceleration of the velocity of expansion was essential to the survival of the steady-state theory, but it appeared to be merely an additional fact for the big bang theory, that could be incorporated by a further assumption. I feel sure that the steady-state theory will regain the center of attention once more, and it is only then that a decision can be made.

Another exciting episode had come from the investigations by Bondi and myself, concerning the damping of the free nutation of the earth's axis of spin, the small angle by which the axis of rotation of the earth differs at any time from the position of the nominal pole. It amounts to movements on the surface of the earth that shift the actual pole by distances of just a few feet from the position of the nominal or average pole position. This had led me to an associated subject, namely, the question of whether the earth's spin could have changed over geologic times, but by a very large angle relative to the earth itself.

Could the poles once have been where the equator is today? It turned out that this subject had been treated in detail by Lord Kelvin and Sir George Darwin (the son of Charles Darwin), the two most prominent earth scientists of the day. They had come to the conclusion that it would not be possible for a major change to have taken place. I had read their lengthy papers on the subject but disagreed with them, considering that they had not treated correctly the three axes that at any time would be almost coincident. Those are the axis of the principal moment of inertia, the axis of spin, and the axis of figure. George Darwin and Lord Kelvin had already been aware of the same values of the plasticity of the earth that I had at this stage, so all our considerations were based on exactly the same data.

It was not easy for me to write a paper challenging these two eminent scientists, and I also doubted whether I could get it published. Nevertheless, I wrote the paper, and to my surprise, the journal Nature accepted it readily. Even better than that, I was shielded from any further debate on the subject because Dr. Walter Munk, a highly recognized geoscientist, followed my paper shortly with one entitled

"A Comedy of Errors," and he regarded my paper as clearly settling the matter. I had concluded that, indeed, the earth could have shifted its poles by large angles, and in recent days, Dr. Kirschvink at Caltech has obtained striking evidence that the earth's position has indeed changed relative to the axis of spin by angles like 90°.

The Strange Case of Io

Io is the closest moon to Jupiter and one of the four moons that were first seen by Galileo. When the NASA Voyager-instrumented planetary research crafts reached the vicinity of Jupiter in 1979 (Voyager 1 in March of that year and Voyager 2 in July), very strange and remarkable effects were observed to occur on Io. Several spots with huge fountains were noted shooting small particles to the great altitude of 270 km. These were immediately declared to be volcanoes. Mountains from which material is expelled at a high velocity are normally called volcanoes when they occur on the earth, and since all the NASA advisors that were concerned with interpretations of the Voyager images were really terrestrial geologists, although they had acquired the title of astrogeologists, they all immediately jumped to the conclusion that the mountains must be volcanoes like their earthlike counterpart. The same had happened in the interpretations of features on the moon, on Mercury, and on Mars. Io was immediately called the most strongly volcanic object in our planetary system.

As we had seen time and again before, it was always the first guess that got all the investigative teams marshaled behind it, and as these groups were mostly led by NASA personnel, it never seemed advantageous for any member to pursue any other line of investigation.

The question arose immediately whether these fountains were permanent features and whether volcanoes could indeed keep eruptions going for prolonged periods, and whether they could propel particles to the heights that were observed.

The second of the two missions arrived at Jupiter 4 months later and the fountains were still seen in much the same places as in the previous mission and also projecting their particles to much the same height. For a terrestrial volcanic event, this would certainly not be normal so that the similarity to terrestrial volcanoes disappeared with that one important observation.

I was, from the start of these observations, quite convinced that they had a totally different origin from volcanism and, in fact, I had explained to many colleagues that in my view, Io would look like a "beat up cathode of a transmitter tube," a phenomenon I knew quite well from my experience of designing high-powered transmitters.

We knew by then that Jupiter had a very strong magnetic field, estimated at about 12 gauss at the surface, and we also knew that radio noise came from the direction of Jupiter, regularly correlated in intensity with the motion of Io around Jupiter. Electrical phenomena were strongly indicated by all this; Io moving in Jupiter's strong magnetic field will make the combination a very effective dynamo, inducing strong electric currents all the time. This would make the near permanence of the fountains much more understandable than volcanic eruptions.

A dynamo is any object that has a conductor moving through the magnetic field of a body and that has electrical connections between the body and the moving object. In our case, the moving object is of course Io. Jupiter itself is the source of the magnetic field and the connections between the two are Jupiter's magnetic field lines. Electric currents can readily flow along those, if the local density is low. It is then a further requirement for the dynamo to work, to have a current flow across the field lines both at Io and in the vicinity of Jupiter. Jupiter will have an ionosphere which, like ours on the earth, will have a level at which the density will be sufficient to conduct electric currents across the field lines.

For Io, it is a different story. If Io were just a lump of cold rock, then its conductivity would not be sufficient to make this arrangement an efficient dynamo. However, it was discovered by Stan Peale, an ex-graduate student of mine, that Io would have a substantial source of heat in its interior as a consequence of the tidal deformation it suffers constantly as its shape accommodates to the gravitational forces exerted by the other satellites. Its orbit represents a resonance with two of the other satellites, and as Io is firmly held in that resonance by the two more massive satellites, it will continuously derive energy for the deformation from the kinetic energy of these other bodies. We had recognized such resonant situations for other bodies in the solar system, but no other case had been quite as extreme as that of Io, for the amount of energy that was made available by this. Peale thought immediately that his discovery had solved the volcano problem because the amount of energy so derived in Io's interior seemed to be sufficient for the energy that was poured out of the several vents.

But this still left the two problems unanswered. Firstly, how can the diffusely distributed energy that was available only in the rocks that were being deformed by the tides all become available at just a few spots? In fact, most of it appeared in just two spots. If the energy was liberated diffusely only in solid rock, it would then have to have been transferred into liquid rock to create the volcanoes. How then could it be made to appear constantly underneath the vents? As liquid rock would suffer very little heating from the tidal deformations, its temperature could never be much higher than the melting point of the solids from which it came, and that, as it turned out, is insufficient by far for the propulsion of particles to the heights that were seen.

But Stan Peale's discovery of the internal heat in Io was still very important for the entire explanation. It was clear that Io's interior heat would make the inside quite hot but leave the outer skin almost as cold as it would have been without the tidal heating. Most rocks belong to the class of semiconductors whose electrical conductivity increases steeply with increasing temperature. This will mean that the interior will be a very good electrical conductor, in contrast to the cool outer layers. If the outer layers contained any cracks or perhaps a genuine volcanic hot spot, then this would become a conductor to allow the electric current to reach the interior, and a corresponding second such spot will be required to allow it to emerge to the exterior again. The deformation of the field lines had been observed by the space craft when it passed between Jupiter and Io, and from this, an estimate could be made of the strength of the current that was in fact flowing along these field lines.

This estimate came to the formidable current of five million amperes. This meant that the energy delivered to Io by the electrical effects was quite comparable with the energy delivered by the tidal effects noted by Peale. Only the body of the mass of Io could become the component of the dynamo that would cause the large current flows.

But now, this energy is automatically funneled into hot spots on the surface, which may be hot spots created by other means, or they may be hot spots directly created by the five million ampere current.

While in metals the conductivity decreases with increasing temperature and, therefore, electric currents tend to spread, for semiconductors like rocks, just the opposite applies. Currents will always tend to funnel down to the location at which they can produce the highest temperature, and in the cold skin of Io, they would, therefore, tend to produce or maintain small high-temperature holes. In this case, the problem of the concentration of the energy to a few spots is solved. The total energy is much the same as in Peale's diffuse heating, but it automatically will tend to concentrate to a few spots. The magnitude of the current observed and the knowledge of the strength of the magnetic field allow the total energy obtained by the dynamo to be evaluated. This rules out the possibility that electric pathways outside the satellite could be making the connections, such as an ionosphere of Io, since the small gravitational force of the external gas would be quite insufficient to extract the required energy from the system. The currents must flow through and bear on the body of Io itself.

Then we come to the problem of the height to which the particles have been seen. Two hundred and seventy kilometers in Io's gravitational field would require that they were given a velocity of about 1 km a second. If this were to represent the velocity of a gas being emitted, we can relate this to the gases known to be present there. Those all seem to be sulfur gases, such as sulfur vapor which is S_2 or sulfur dioxide which is SO_2 or hydrogen sulfide which is H_2S. We can then calculate what temperature each one of those gases would require to create a fountain that reaches the observed heights. The propulsion of particles would have to depend on gases that have reached such velocities and carried the particles to a common velocity. This would be in most cases less than the speed of sound in the gas, and so we can derive the temperatures required. While in the volcanic explanation one could never assume a higher temperature for the gas than the melting point temperature of the confining rocks, in the electrical case, there is no such limitation. It is well known an electric arc can be struck which will burn a hole and vaporize the material. The temperature is not even limited by the temperature of vaporization of the rock, since the electric current will continue to heat gases that are not in contact with the rock anymore.

It was very clearly seen, at least in the case of one of the prominent plumes, that it had, quite accurately, the shape of an umbrella. This means it was not a diffuse distribution of velocities of individual particles, but they were initially driven to a common velocity and a common height, just as is the case for a water fountain which then splits up at the pinnacle and spreads apart. A gas fountain would have a similar property, and therefore, if any particles condensed from the gas as it cools, they would all start out with the common velocity that the gas had achieved.

For sulfur or sulfur dioxide, that velocity would require an initial temperature to be in the neighborhood of 6,000 K, while for hydrogen sulfide, it would be only about 3,000 K. All these temperatures are much too high when compared with the melting points of rocks. This had indeed been a point of contention between me, with my electrical theory which could reach these high temperatures, and the NASA group of astrogeologists who knew that they would be limited to a temperature of the order of the melting points of rocks. Their reply to my paper in Science on the electrical origin of the outbursts on Io was that the high temperatures I mentioned had not been observed. My personal response to that was "you may not observe the high temperatures in the furnaces of power stations when you fly over them." I received many favorable responses to my letter in Science, but Science did not allow me to break through the solid barrier put up by NASA.

As a result of the Galileo Mission in 2003, the situation changed substantially. Temperatures in the ranges that I had indicated have indeed now been seen, and the people who favor the volcanic interpretation are engaged in looking around whether perhaps some unusual rocks exist that have such high melting points that a volcanic interpretation can still be maintained. Science has not invited me to submit another paper now that the high temperatures have been seen, although their editors had previously regarded the absence of observations of the high temperatures to be signaling the end of my electrical explanation.

Pulsars

The episode of the discovery of pulsars was plain sailing for me. I had previously made the prediction that condensed stars could act as powerful radio beacons, but would then be expected to show time variations in their radiation comparable with the flight time of light across their dimensions. When I made that prediction, in 1952, no such objects were known. When they were discovered in 1969, I had to do no more than go back to my earlier considerations which were in print and fitted very well with the new discoveries.

I had taken to publishing immediately a number of predictions concerning future results of observations predicted by new theories. For pulsars, this included the prediction that the sites of comparatively recent supernovae explosions would be particularly favorable for finding more pulsars and that they may be very much spinning faster and thus showing much higher pulse repetition frequencies than were seen in the first pulsars discovered. Also, they would be found to have the pulse repetition frequencies slowing down slightly. All this was regarded at the time as "far-out stuff." But it did not take long before the Australians discovered a faster pulsar in a known supernova site (Vela) in the southern hemisphere, quickly followed by the discovery of an even faster one in the location of the best known recent supernova, the Crab Nebula, which showed also the predicted slowdown of the repetition frequency. With that, the chapter was closed. To make predictions for any new theory was mainly necessary to avoid the criticism that perhaps the right answer had just been a shot in the dark.

But how did predictions fair in the other cases?

The Swedish Drilling Experiment

For the theory of hearing, it took about 30 years before the predictions were recognized and identified. It was of course, for me, a disappointingly long time. For the theory of large polar wonder over geologic times, it was an even longer interval before anyone looked for clues, an interval of about 50 years. In this case, I had not expected anything much faster since I knew of no one who was even working on this. But for the great experiment with the deep bore holes in Sweden, the situation for me was quite different. Since I was seeking to collect large sums of money for the drilling process, I had to explain that the chances of finding oil and gas in Sweden were pretty good, but I had to make sure not to lay myself open to the accusations that I was deliberately misleading people since in the financial matters of this magnitude, that would be considered a crime.

Most of my public speeches in Sweden about this subject started out with the words "It is unusual in oil and gas exploration to have commercial success with fewer than five holes. It just takes that long before all the difficulties that arise in the new domain can be ironed out, so I suggest that a plan be made and finances collected that would allow five deep holes to be drilled in the granite of Sweden." In fact, that was never achieved, and drilling started when there was only enough capital for a single hole. Eventually because of the great promise shown in that first hole, more private money was raised to drill a second one.

During the drilling of the second hole, we had extremely strong favorable information, namely, that several substances usually associated with the most successful well bores were present in great abundance. So then, the testing procedure became a cause of great excitement. Telefaxes were sent to me every morning, and so every morning brought the excitement of the possibility that a major find had been made and a large industrial country like Sweden might have been made independent of oil imports. The magnitude of the formation into which we were drilling was so large, and the favorable indications extended over the entire region, so that one will have been entitled to say that an oil field the size of that of Kuwait lay underneath. There could be no question of the amount of oil that was there because that would have been the amount of oil that one could actually obtain from the drilling chips that were dilled up. It was therefore only a question not of the presence of the oil but of the ability to draw it out of the rock. When eventually a down hole pump produced eighty barrels of oil in 2 days, we thought that we were very close to our aim. The oil was tested and found to be perfectly ordinary, natural crude oil, and the only thing that was peculiar about it was that it came into the well-bore mixed with very fine-grained magnetite particles which later turned out to be bacterially produced. That of course was the discovery that implied the existence of a very large amount of living material, which itself was fed mainly by hydrocarbons that came up constantly from below.

While that in itself was a great discovery, it was the same action that in the end prevented commercial exploitation of the well. The porous surface of the well-bore kept being blocked up by the small particles, and so every time it had produced a few barrels of oil, it stopped flowing. Ironical, but sad, that the bacterial action on which we had counted to supply all petroleum with biological molecules, later referred to as "the deep hot biosphere," was just the process that prevented a commercial success.

Acknowledgments

There are many other people, in addition to those that I have mentioned in the text, who have had a major effect on my scientific career and on my thoughts.

I have mentioned Fred Hoyle who introduced me to many problems of astronomy and with whom I discussed many thoughts and ideas that had occurred to me. He remembered them all very carefully and wrote about them in his prolific publications. Also Richard Pumphrey had a great influence by asking me to solve many riddles he had come across in his thoughtful scientific life.

Then there was Sir Harold Spencer Jones, the Astronomer Royal, who brought me to the Royal Observatory as his chief assistant, and with whom I also had many scientific discussions during my time at the observatory. Arthur Kantrowitz, the high-speed gas dynamics expert, took up many of my ideas and went to great lengths to show that my notion of MHD shock waves was indeed correct and an important item in the discussion of high-velocity gas dynamics.

It was Donald Menzel, the Director of the Harvard College Observatory, who invited me to Harvard and to a professorship there. Dennis Sciama, a longtime colleague of mine, was also a person who demanded detailed explanations and justifications of many topics I picked up over the years. At Cornell, the spirited attitude toward science of Phil Morrison was always a great encouragement.

I mentioned some high spots of discussions with Richard Feynman which also increased my self-confidence, but which mainly showed me how much faster a human brain can work than mine could.

Harold Urey almost always very favorably inclined to my ideas, always a great pleasure to visit, and always a good personal friend. I mentioned Werner Heisenberg who was also a delight in scientific discussions because he understood everything so quickly and I enjoyed his replies.

Steven Soter has been my constant companion over many of the years I have described here, and, with great care, he went through many of the calculations and arguments I presented. Often, he forced me to find better arguments when mine were not clear enough or inadequate. He was unquestionably the most important and best collaborator I have had.

Thomas Gold *edited by* Simon Mitton, *Taking the Back off the Watch*,
Astrophysics and Space Science Library 381, DOI 10.1007/978-3-642-27588-3,

The correspondence with many Russian geoscientists and geochemists, chiefly Peter Kropotkin, was an important item in my development of ideas about the origin of petroleum and other items relating to the deep subsurface of the earth. But, I am sure there must be many others who I have now failed to recall.

For the preparation of the present autobiography, I want to thank the Richard Lounsbery Foundation for providing me with secretarial help when I urgently needed that to complete the book and for the help of my secretary, Claire Perez, who would not only type my dictated text but proofread it and find the required references.

Although it was a lot of work to dig back that far into my past and to try and remember the major occurrences, I found it a pleasure to go over all this once more as if I had a chance to relive my scientific life again.

A Personal Postscript

Tommy had a very gentle side that many of his professional colleagues did not see. He loved music, animals, nature, and his family.

He also loved to cook. He used to joke that he thought the reason I married him was because he was a good cook. He taught me the best of Austrian cooking, from Wienerschnitzel to Apfelstrudel.

I think that Mozart was his favorite composer, and he used to love to sing along to some of the operas, The Marriage of Figaro being his favorite. He knew a little Italian and would translate the songs as they were being sung, over and over, until I too knew them by heart.

His pets were a beloved part of his family. He made a little cart for his Samoyed Tasha, which, to his children's delight, the dog would pull around with them in it. When he was retired and working at home, he was heartbroken when his beloved golden retriever Kim passed away, saying that he would miss talking to her during the day while I was working. He was emotionally unable to go with me to the veterinary hospital to say good-bye to her, but just sat quietly in his chair at the kitchen table until I returned several hours later. We then went out for a long walk, holding hands and gaining comfort from each other.

He wrote poetry for his children when they were young, which I collected into an album that I called "The Golden Book of Verse." For the most part, they were humorous little verses which he loved to read to his children, and they loved to recite back to him, alternating verses, with Tommy reciting one line and they the next.

When Tommy was gravely ill with very little time left, one of his children brought a favorite book of children's verse to the hospital. Tommy was barely able to speak at that point, but Tanya began to read a verse from the book. A smile came to his face, and a tear to Tanya's eye, as he struggled to recite the next line back to her. They finished the whole poem.

Tommy was a man of many different colors to many different people. I knew them all, since I was his assistant at Cornell and knew all of his colleagues personally as well. But it was his gentle side that I miss the most.

September 2007 Carrie Gold

Thomas Gold *edited by* Simon Mitton, *Taking the Back off the Watch*,
Astrophysics and Space Science Library 381, DOI 10.1007/978-3-642-27588-3,

Index

Thomas Gold *edited by* Simon Mitton, *Taking the Back off the Watch*, Astrophysics and Space Science Library 381, DOI 10.1007/978-3-642-27588-3,

Astrophysics and Space Science Library

For further volumes:
http://www.springer.com/series/5664

The manufacturer's authorised representative in the EU is Springer Nature Customer Service Centre GmbH, Europaplatz 3, 69115 Heidelberg, Germany. If you have any concerns regarding our products, please contact ProductSafety@springernature.com

Printed and bound by CPI Group (UK) Ltd, Croydon, CR0 4YY
30/09/2026
02225595-0014